LIBRARY
WARWICKSHIRE
COLLEGE

Warwickshire College
* 0 0 5 1 0 1 7 9 *

HORSE GENETICS

To my father

Horse Genetics

Ann T. Bowling

Veterinary Genetics Laboratory
School of Veterinary Medicine
University of California
Davis, California, USA

CAB INTERNATIONAL

CAB INTERNATIONAL
Wallingford
Oxon OX10 8DE
UK

Tel: +44 (0)1491 832111
Fax: +44 (0)1491 833508
E-mail: cabi@cabi.org
Telex: 847964 (COMAGG G)

A catalogue record for this book is available from the British Library.

ISBN 0 85199 101 7

Typeset by Techset Composition Ltd, Salisbury
Printed and bound in the UK at the University Press, Cambridge

CONTENTS

PREFACE

Basic genetics textbooks seldom mention horses. Horse breeders tell me that they cannot relate fruit flies, corn and mice to practical horse breeding. This book aims to provide a good overview of genetic principles using horses as the primary examples. I have sifted and distilled facts and ideas to provide horse breeders with relevant illustrations. While providing a basic primer, I will not oversimplify to the point of inaccuracy. Students and science professionals can confidently use this handbook as a resource.

Do not expect to read this book from cover to cover! In my years of teaching, I have found that genetics is a subject that can only be taken in small doses. When you reach your saturation point, stop for the moment to return at another time.

Browse the contents, the pictures and tables to become familiar with the material present. You may want to read only selected chapters or sections. The index may point you to several sections discussing a subject of particular interest. If you wish to read the original research papers, the reference section provides the citations to find them in a university library. You will probably want to have a general genetics textbook at hand to refresh your memory or to provide alternative and more detailed examples of basic principles.

Knowledge about horse genes lags well behind that for human or mouse, or even for genes of other domestic animals such as cow, pig, sheep and chicken. Since the horse provides only a very limited set of examples, readers keen to know more about genetics are encouraged to consult current texts in general genetics. I especially recommend the veterinary genetics textbook by Nicholas (1987) for its wealth of animal examples.

You may find the contents overwhelmingly detailed about Paint horse pattern genes. The emphasis is a reflection of the many inquiries I receive from Paint breeders. Even if white spotting genes don't pertain to your breeding program, they provide examples to help you practice thinking about horse genetics. You may be disappointed that no discussion is provided on a particular subject important to your breeding program. Information specific to horse genetics comes at a price. Government-funded agriculture programs support research on food and fiber animals, not companion animals. Knowledge about horse genetics will be forthcoming in direct proportion to how much money is invested. Without money being committed to horse research by horse breeders, our understanding about genetics of humans, mice and cattle will continue to advance rapidly, but knowledge about horse genetics will only unfold slowly.

Readers of this book will find answers to many of their questions about the genetics of horses, but I hope that other questions will replace them. Learning is a continuous process that does not end with finding answers. The path of knowledge is learning to ask questions and to build new questions from the answers.

Ann T. Bowling

ACKNOWLEDGEMENTS

This book is the product of my intellectual and personal experiences. Thousands of people have affected this effort, but it is the contacts of the last few years that are naturally to the fore of my memory. The book is based on a solid base of factual data provided by my colleagues and students at the University of California, Davis; national and international colleagues in blood typing and cytogenetics of horses; and researchers not personally known to me who have studied and published information about horse genetics. Students and horse breeders who ask good questions have made an important impact on this book by helping me to choose content that provides them with a solid foundation in basic genetics as applied to the horse.

Particular gratitude and thanks are owed to my husband, Michael, also a geneticist specializing in horses, who helped me write what I wanted to say when my choice of words failed me. He also patiently survived weekend physical and mental absences while I pursued this project. Thanks also to our daughter Lydia, who endured a mother at times mentally preoccupied with academic pursuits, not cooking, reading and shopping for school clothes. This book is not my effort alone, and to all who contributed directly and indirectly, I am sincerely grateful.

ILLUSTRATION CREDITS

The author is grateful to friends and colleagues who provided figures and photographs for this book. Thanks and credit are due for the following figures (other figures and photographs are from the author's collection, except where noted in the figure caption):

Credit	Figure
Alison Schafer	45
Barbara Naviaux	16, 41, 42
Dean Neely, DVM	47
Dixie Morrow	21
Equine Research Laboratory, UCD	31
Peter and Deborah Bowling	50
Leon V. Millon	2
Linda Sue Nickel	10, 60
Michael Bowling	11, 44
Waltraut Zimmermann	15

ABBREVIATIONS

21-OH	Steroid 21-hydroxylase
A	A blood group
A	Agouti
A1B	A-1-B-glycoprotein
ADA	Adenosine deaminase
AHG	Anti-hemophilic globulin
AHRA	Arabian Horse Registry of America
ALB	Albumin
AMHA	American Morgan Horse Association
AP	Acid phosphatase
APHA	American Paint Horse Association
AQHA	American Quarter Horse Association
BLUP	Best linear unbiased prediction
C	C blood group
C	Color
C3	Complement component 3
C4	Complement component 4
CA	Carbonic anhydrase
CAT	Catalase
CID	Combined immunodeficiency disease
cM	Centimorgan
CRC	Calcium release channel
"*D*"	Nei's distance
D	D blood group
D	Dun
DIA	NADH-diaphorase
DNA	Deoxyribonucleic acid
Dom	Dominant megacolon (mouse)
DRA	DR-α (a gene in the MHC)
E	Extension
ELA	Equine lymphocyte antigen
ENB	Equine night blindness
ES	Carboxylesterase
ESCI	Equine soluble class I antigen
F	Inbreeding coefficient
FISH	Fluorescence *in situ* hybridization (physical gene mapping)
G	Gray
GC	Group-specific component (vitamin D binding protein)
GOTm	Glutamate oxaloacetate transaminase, mitochondrial
GPI	Glucose phosphate isomerase

h^2	Heritability
HBA	Hemoglobin-α
HLA	Human lymphocyte antigen
HME	Hereditary multiple exostosis
HMS	A series of microsatellite loci (DNA)
HP	Haptoglobin
HTG	A series of microsatellite loci (DNA)
HYPP	Hyperkalemic periodic paralysis
H–W	Hardy–Weinberg law (population genetics)
IDH2	Isocitrate dehydrogenase 2
IGF-1	Insulin dependent growth factor-1
ISAG	International Society for Animal Genetics
K	K blood group
KIT	Tyrosine kinase transmembrane receptor (mouse)
LDHB	Lactate dehydrogenase B
LG	Linkage group
LP	Leopard spotting
ls	Lethal spotting (mouse)
LWFS	Lethal white foal syndrome
ME1	Malic enzyme
MHC	Major histocompatibility complex
mi	Microphthalmia (mouse)
MPI	Mannose phosphate isomerase
mtDNA	Mitochondrial DNA
NI	Neonatal isoerythrolysis
NOR	Nucleolar organizer region
NP	Nucleoside phosphorylase
O	Overo
OAAM	Occipital-atlanto-axial malformation
OCD	Osteochondritis dissecans
P	P blood group
PCR	Polymerase chain reaction
PEPA	Peptidase A
PEPB	Peptidase B
PEPC	Peptidase C
PGD	6-phosphogluconate dehydrogenase
PE	Probability of exclusion
PGM	Phosphoglucomutase
Ph	Patch (mouse)
PI	Protease inhibitor
PLG	Plasminogen
PtHAA	Pinto Horse Association of America
QH	Quarter Horse
QTL	Quantitative trait locus
RBC	Red blood cell
RFLP	Restriction fragment length polymorphism
RN	Roan
Rw	Rump-white (mouse)
s	Piebald (mouse)

SCID	Severe combined immunodeficiency disease
SLR	Simple length repeat (DNA)
SNPs	Single nucleotide polymorphisms
Sp	Splotch (mouse)
STR	Short tandem repeat (DNA)
STS	Sequence tagged site (DNA)
TB	Thoroughbred
TF	Transferrin
TNFA	Tumor necrosis factor
TO	Tobiano
U	U blood group
VHL20	A microsatellite locus (DNA)
VNTR	Variable number of tandem repeats (DNA)
W	White
Z	Silver dapple

CHAPTER 1
Basic genetics

Working with genetics can be fun and rewarding for those people who have logical thinking skills and enjoy playing sleuth. Many horse breeders will enthusiastically take on the task of learning about genetics, especially for the breed they have chosen, but interest in horse genetics is not limited to people with a breeding program. Buyers, even if not planning to breed horses, need to know what may and may not be reasonably expected. Horse owners may be fascinated to learn how one favorite steed comes to be distinctive in color, size, shape or ability from another.

A horse show provides a good opportunity to compare aspects of breed type that identify individual horses as belonging to a particular breed. The most conspicuous differences include size, gait and carriage, but shape of head, neck and croup may also be distinctive. Within each breed, individual horses provide another spectrum of differences, particularly for color and markings, but also with their closeness to the ideal breed type. Distinctive features of type, color, markings and way-of-going are inherited traits conditioned by the actions of particular sets of genes. However, despite our observation of differences, comparing genes between breeds would show that nearly all the genetic information is identical. After all, we are comparing *horses*, not animals of different species.

In theory, the similarities should make it a relatively easy task to understand the genetic differences between breeds or individuals. Practically speaking, thousands of genes need sorting out. For the most part we know very little about the specific genes responsible for distinctive traits. Lack of objective components to score or measure makes defining genes a slow process, but new genetic technologies will provide more effective tools than have been available.

In 1866 the Austrian monk Gregor Mendel first clearly described the principles of genetics from his work with garden peas. Mendelian genetic principles apply to inheritance of traits in animals as well as plants. For both animals and plants, it is always critically important to recognize the environmental influence on an organism's characteristics. This book will help a horse breeder learn not only *how* traits are inherited but also *which* traits are inherited.

The aim of this first chapter is to describe the basic principles of Mendelian genetics using horses, not peas, as examples.

What are genes?

Genes cannot be seen, even with the help of a microscope, but it is not necessary to see a gene to predict the outcome of matings. Mendel never saw a gene, yet he was able to describe the basic principles of genetics. To students of horse genetics today, as to Mendel, genes are known primarily through their effects (e.g. black or chestnut hair color).

Genes are units of inheritance

Genes are passed from parent to offspring through hundreds of generations, essentially *without modification*. Why then don't offspring always look like their parents? Differences between generations result from *new combinations* of genes in an offspring compared to its parents.

Successful livestock breeders, either through intuition or formal study of genetics, use gene combinations to their advantage. Breeders with excellent stock may want to have as few differences between parental pairs as possible, because new gene combinations may not be desirable for their breeding goals. Breeders with stock that do not meet breed or production standards may combine gene differences between parents to produce offspring that exceed the parental norms.

Genes are linear arrays of nucleotide sequences

Genes are very large, complex molecules of **DNA** (deoxyribonucleic acid). To direct the incredibly complicated life processes of each organism requires an estimated 50,000–100,000 genes. Despite the complexity, an elegant simplicity characterizes DNA. DNA is composed of four **nucleotide** units in linear arrays of great length. The nucleotides differ from each other according to which base they contain: **A** (adenine), **T** (thymine), **G** (guanine) or **C** (cytosine). They can potentially form an endless number of combinations, but the sequences are not composed at random. Nucleotide sequences are *the code of genetics*.

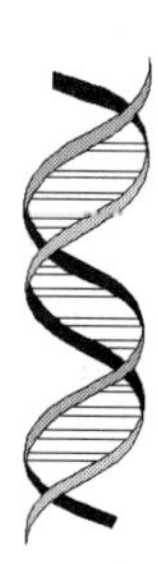

Figure 1: The coiled strands of the DNA double helix are held together by bonds—like rungs of a ladder—between paired nucleotides in the linear arrays. The strands have complementary sequences and when the cell divides each half receives identical genetic information.

The physical structure of DNA is a **double helix** (Figure 1), a key component of genetics whose discovery by James Watson and Francis Crick won them the 1962 Nobel Prize. The double helix is composed of complementary strands—for every A in one strand, a T is present in the other and for every C, a G. Helical structures coil and supercoil, allowing alternative phases of DNA extension and condensation critically important for gene activity and cell division. The code is passed on to every new cell through a mechanism that uses each strand as a template to generate the complementary strand. In this way when the cell divides, each product receives an exact copy of the code.

Some genes are units of information for production of proteins

For protein synthesis, the code is read from only one strand of the DNA double helix. A very tiny part of the sequence for the horse muscle sodium channel gene is: ATCTTCGACTTC. This order encodes information for assembling **amino acids** into a **protein**. Reading the code in groups of three letters, these 12 bases are translated into a string of four amino acids (isoleucine–phenylalanine–asparagine–phenylalanine), a very small piece of a large protein molecule that is a part of each muscle cell membrane.

Proteins are substances that control the developmental steps from ovum and sperm through foal growth stages to adult. Proteins determine form and structure, and provide information to maintain life functions. Extremely rarely, a change in the DNA sequence (**mutation**) occurs that may alter the structure of the protein produced (or result in no protein), causing early embryonic death or an inherited disease. In our muscle sodium channel example, a change of the second C to a G results in a substitution of the amino acid leucine for phenylalanine and is associated with hyperkalemic periodic paralysis (HYPP), an inherited muscle paralysis disease of Quarter Horses and breeds that use Quarter Horses as breeding stock (such as Appaloosas and Paints). Not all changes to DNA result in visible consequences to the animal. Some mutations alter nucleotide sequence but do not cause an amino acid substitution and thus do not have deleterious or otherwise recognizable effects (**silent mutations**). Silent mutations may also occur in regions of DNA that do not code for a protein.

Some genes are units controlling biological information systems

All an animal's cells contain the same genes but specific genes in each cell need to be "turned on" and others "turned off." For example, genes to control bone production are turned off in liver, but are active in bone growth plates. Less than 10% of human DNA is in **exons**, the regions that code for protein products. One function of the remainder, the **introns**, appears to be controlling the behavior of genes.

Although their function for the organism is presently unknown, throughout the genetic material are runs of repeating nucleotide units. Large, complex repeats are commonly called **minisatellites**. They are used in so-called fingerprinting procedures. Runs of simple nucleotide repeats called **microsatellites** or STRs (short tandem repeats) (e.g. CACACACACA) are distinctly different from protein coding regions for which the largely unique sequences (genes) will be composed in each DNA strand of all four nucleotide components. With the wealth of inherited variation that can be studied with newly emerging DNA technology, STRs are proving to be effective tools for parentage studies and gene mapping.

Gene information may be similar to that of humans and other animals

Comparisons between genes of horses and other animals show great similarities. Horses will indirectly benefit from the Human Genome Project, which proposes to define the DNA sequence for the approximately three billion nucleotides in the human genome. As gene sequencing technology becomes more automated and less costly, the number of genes defined for horses is expected to increase substantially. Why is horse gene sequence information important to horse breeders? Genome information will lead to new possibilities for understanding, diagnosing and predicting genetic traits. If genetic diseases in horses have well-studied counterparts in humans, gene carrier assays may be more rapidly developed than if the horse research had no candidate genes as a starting point.

Where are genes found?

Genes in the nucleus

The DNA of horse genes is packaged in 64 **chromosomes** found in the **nucleus** of every cell. Chromosomes can be seen with the aid of a microscope and dyes that bind to DNA. The genetic information of all horses is nearly identical and, not surprisingly, horses of all breeds have the same number, size and shape of chromosomes. Crosses between breeds widely different in appearance readily produce fertile offspring, providing additional evidence of the basic genetic similarity of breeds.

When the cell starts the process of division into two daughter cells, the chromosomes condense by super-coiling from their extended state resembling tangled spaghetti into discrete rod-shaped bodies (Figure 2). Careful cutting and matching of stained chromosome images obtained from an enlarged photographic print of a cell in the process of division shows that the 64 chromosomes can be arranged as a series of 32 pairs of structures. The array of paired chromosomes is known as a **karyotype**. The only distinguishing feature between most horse karyotypes is a difference between males and females seen in a single pair of chromosomes discussed in a later section.

Genes in mitochondria

It is the genes in the nucleus that follow Mendelian genetic principles. A few genes are found in cells, but outside the nucleus, in structures called **mitochondria**. These genes are also composed of DNA but have a different pattern of inheritance than nuclear genes.

The behavior of chromosomes

The conduct of chromosomes through cell life cycles is the key to the principles of Mendelian inheritance. Two types of division cycles are characteristic

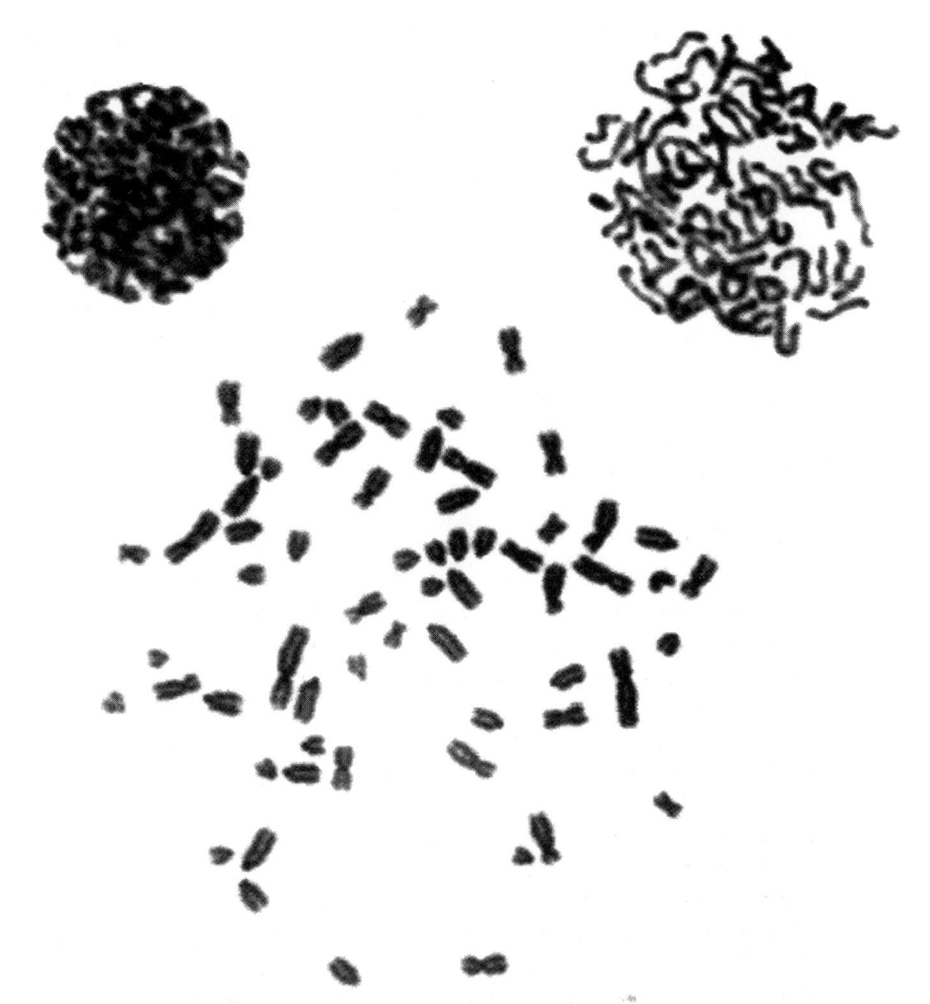

Figure 2: Microscopic images of dye stained nuclei from three horse lymphocytes (white blood cells) undergoing cell division show progressive phases of chromosome condensation, from a tightly coiled mass (upper left) to a chromosome spread (bottom) in which each of the 64 elements is individually distinguishable. A paired array of chromosomes known as a karyotype could be prepared by cutting out images from the photograph of this spread (see Chapter 12).

of chromosomes. The first process (**mitosis**) occurs in all cells of the body. The second chromosome process (**meiosis**) is directly involved in formation of the gametes (reproductive cells) and occurs only in the reproductive organs or gonads (testes in males and ovaries in females).

Mitosis

When body cells divide, the chromosomes first precisely duplicate themselves, then condense by tight coiling to become the discrete chromosome elements shown in a karyotype. At cell partition, the duplicated strands separate so each daughter cell has an exact replica of the genetic material of the original cell. This process assures that all cells of the body are genetically identical and have the normal chromosome number (the **diploid** number). For domestic horses the diploid chromosome number is 64, a collection of 32 pairs of chromosomes. One chromosome of each pair has a maternal origin, the other a paternal. Rarely, mutational events may occur in the DNA of body cells, but the changes cannot be passed on to a subsequent generation unless they occur in reproductive cells.

Meiosis

Meiosis generates **gametes** (sperm in males and ova in females) with only 32 chromosomes (the **haploid** number)—only one copy from each of the chromosome pairs found in normal diploid cells. When a sperm and an ovum combine during fertilization to form a **zygote**, the chromosome number in the resulting cell is 64, reconstituting the chromosome number and gene composition appropriate for the animal we know as the horse.

Integral to meiosis are two programs that are directly responsible for the characteristics of gene inheritance.

- **Reduction division** results in the gamete's receiving only one chromosome of each pair, randomly distributing parental-derived chromosomes through their children on to the grandchildren. This process reassorts chromosome pairs in each generation and generates characteristic trait ratios and segregation of alleles. Mendel did not know about chromosomes but he hypothesized this kind of process to explain inheritance.
- **Recombination** allows homologous maternal- and paternal-derived chromosomes to exchange sections. This crossing-over process was not part of the genetic theory hypothesized by Mendel but is the basis for the important concept of linkage genetics.

An animal has only two copies of each gene despite the genetic input from many pedigree elements. For example, all four grandparents will provide material to the overall genetic makeup of a grandchild, although for each specific gene only two grandparents, one from the paternal side and one from the maternal side, will be represented. Certain groups of genes are likely to be co-contributed because genes are closely strung together on linear chromosomes. Meiosis ensures that genes on different chromosomes or far apart on one chromosome are unlikely to stay together, even through only a few generations.

If this brief summary of cell division processes is not sufficient, consult a basic text on genetics for a more detailed review. For this topic, it would make very little difference for understanding the fundamental process whether a mouse, a fly or a horse was the example. From the description of the various cell and chromosome division processes, and basic to all that follows, the key point to understand is that individual genes—the units of heredity—are passed on unaltered from parent to offspring, but *the gene combinations are changed in every generation*.

The inheritance of sex

A foal's sex is determined by the genetic contribution it receives from its sire, not its dam. No other trait is known to be specifically determined by the stallion, but we will spend time on sex determination since it provides an uncomplicated example of the way that gene (trait) transmission directly follows the behavior of meiotic chromosomes.

A clear difference between the karyotypes of males and females can be seen in one pair of chromosomes, called the **sex chromosomes**. In the male this pair has different elements, while in the female these chromosomes are indistinguishable. The sex chromosomes of the male are designated XY. The sex chromosomes in the female are both like the X of the male and thus her pair is designated XX. The other chromosomes, apart from those involved in sex determination, are called **autosomes**. The members of every chromosome pair are split up during gamete formation. Every gamete receives 31 autosomes and one sex chromosome. All gametes of the female have a single X chromosome. Male gametes are an equally divided mixture of X- and Y-bearing sperm. Any ovum fertilized by an X-bearing sperm results in a filly. A Y-bearing sperm produces a colt.

Since the sperm is equally likely to contain either an X or a Y chromosome, female and male offspring are equally likely to occur. For each offspring produced, the chance of being male or female is 50%. *The sex of each offspring is independent of the sex of any previous offspring.*

Geneticists often use a simple diagram called a checkerboard or **Punnett square** to predict the outcome of matings. On the top of the diagram are listed the alternative traits contributed by one parent (in this case the stallion's chromosomes, symbolized X and Y); at the left are the alternatives from the other parent (in this case the mare, who contributes only a single X trait).

Genetic contribution from ova	Factors contributed by sperm	
	X	*Y*
X	*XX* female	*XY* male
Offspring proportion	50%	50%

Figure 3: Punnett square shows the expected outcome of sex chromosome distribution from sire and dam to offspring, predicted to produce a 1 : 1 ratio of male to female offspring.

At the intersection of the columns and rows are the factor combinations that can be produced (Figure 3).

Important genetics lessons to learn from the study of sex determination are:

- Equal ratios of the trait alternatives (in this case, sex) are expected among the offspring in a cross of this type. The arithmetic of genetics is that of chance. Genetics is like a game of coin tossing—with hundreds of coins in the air at once. The outcome of each coin toss is an independent event.
- The inheritance of Mendelian traits follows the inheritance pattern of chromosomes. Chromosomes occur in pairs so the genetic information for each gene is present in duplicate, but only one of the two alternatives will be transmitted, at random, from each parent to each offspring.
- Genetic contrasts between siblings could be determined solely by a difference in the contribution of one parent (for other traits, this will not always be the male). When we are evaluating the inheritance of traits, our task may be made easier if we can recognize or set up situations in which trait variation is only determined by one parent (e.g. a test cross, to be discussed later), although for some types of traits this may not be possible.

The language of genetics

Genes and alleles

Occasionally an offspring is distinctly different from either parent. For example, a pair of black horses may produce a red (chestnut or sorrel) foal, as well as the anticipated black ones. Numerous studies have verified that red hair color is inherited as an alternative to black. The alternative states of a particular gene are called **alleles**.

Two gray horses can produce a red foal. Does that mean that red is also an allele of gray? No, the alternative state (allele) for gray color is "not gray." In our example, the "not gray" horse happens to be red.

The proper assignment of traits as allelic alternatives is not always intuitively obvious. Scientists may propose allelic relationships based on information from similar traits that have been described in other animals. Novice geneticists probably will need to accept the given definition of traits as alternative alleles but eventually they will want to understand how allelism is proven through breeding trials.

Alleles may differ from each other by a single nucleotide base of the gene's DNA sequence. The change may be sufficient to alter the function of the gene product and create a new visible trait. For most alleles of horse genes the differences have not been described at the molecular level, so at present they are defined in terms of their influence on visible appearance.

Dominant and recessive

When black parents produce a red foal, the red color factors were carried in the genetic stuff of the black parents although they could not be seen by someone looking at the horses. The red allele is said to be **recessive** to black, and the black allele is the **dominant** alternative to red. A dominant allele is expressed even when carried by only one member of the chromosome pair. A recessive allele is expressed only when a dominant alternative is absent. A characteristic that is often useful to know about a recessive trait is that it will always breed true—red bred to red will always produce red.

A dominant allele is not necessarily associated with strength, nor is a recessive a sign of weakness. One may hear the expression "hidden" recessive, with undesirable connotations implied if the recessive trait is considered defective, but certainly not all recessives are deleterious and deleterious genes need not be recessive.

The terms dominant and recessive describe the relationship between alleles of one gene, not the relationship between different genes. This point *must* be understood by anyone interested in predicting genetic traits such as coat colors and will be repeatedly emphasized.

Phenotype and genotype; homozygous and heterozygous

The difference between black and red is due to alleles of extension, the black pigment gene, symbolized as *E*. The dominant allele is assigned a capital letter with a superscript (E^E), which can be simplified for discussion to *E*. The recessive allele is E^e or simply *e*. Assignment of the same symbol to a gene and the shorthand version of its dominant allele may seem inappropriate, but this practice seldom leads to confusion. In describing the genetic traits of any animal, the **genotype** designates the alleles present and the **phenotype** specifies the external appearance resulting from interaction of allelic pairs.

Genotype	Phenotype
EE	black hair
Ee	black hair
ee	red hair

The genotype of a red horse will always be *ee*. Because the black allele is dominant to red, the genotype for a black horse is not readily apparent from the phenotype. Genotypes *EE* and *Ee* will both have a black phenotype (horses may be bay, black or buckskin—other genes control the differences among these colors). By convention, the genotype may be designated as *E–* if the second allele cannot be determined from parent or progeny information. When both alleles in a pair are the same (*EE* or *ee*), the horse is said to be **homozygous**; when the pair has unlike alleles (*Ee*), the horse is **heterozygous**.

Genetic contribution from *Ee* mares	Genetic contribution from *Ee* stallions	
	E	*e*
E	*EE* black 25%	*Ee* black 25%
e	*Ee* black 25%	*ee* red 25%

Figure 4: This Punnett square shows the cross between heterozygous black horses, producing black and red foals in a 3 : 1 ratio. The black producing allele *E* is dominant to the red allele *e*.

A Punnett square shows how black and red foals can result from a mating of two blacks, and the expected proportions (Figure 4).

One way to determine which black foals are *EE* and which are *Ee* is through a **test cross** to an *ee* (red), since no direct assay for the *e* allele is available. A test cross is a mating between a homozygous recessive animal and an animal with the phenotype of the dominant allele. Such a cross provides evidence to define the genotype of an animal with the dominant phenotype. In a test cross, a homozygote (*EE*) for the dominant phenotype will never have foals of the recessive (*ee*) phenotype, but a heterozygote (*Ee*) will have foals of both the dominant and recessive phenotypes in a 1 : 1 ratio.

Sometimes pedigree or family study can help determine genotypes that may not be obvious from phenotype. For example, a black with a chestnut sire, and any black that produces a chestnut foal, will necessarily be *Ee*.

Ratios for lethal traits

White coat color is one of the few lethal genes described for horses. All non-white horses are *ww*. As far as is known no horse is homozygous for *W*. Study of white horse breeding data showed that in repeated matings between white horses, both solid color (non-white) and white foals were always produced (Figure 5). The ratio of white to colored foals (28 : 15) more closely approximated the 2 : 1 ratio anticipated for a gene with a homozygous lethal class (presumably *WW*), than the 3 : 1 ratio anticipated if all offspring were equally viable (Pulos & Hutt 1969).

Incomplete dominance and codominance

Occasionally homozygous and heterozygous genotypes for a dominant trait have recognizably different phenotypes. While this is not the case for *EE* and *Ee*, another coat color provides an example of this effect. A coat color dilution gene of horses can be used to demonstrate **dosage effect** or **incomplete dominance**. Palominos and buckskins are heterozygous for the cream

Genetic contribution from *Ww mares*	Genetic contribution from *Ww* stallions	
	E	*w*
W	*WW* lethal 25%	*Ww* white 25%
w	*Ww* white 25%	*ww* not-white 25%

Figure 5: If all classes are equally viable, the proportion of white to colored offspring from matings between white horses heterozygous for *W* is expected to be 3 white : 1 colored. If one homozygous class is lethal, then the proportion of white to colored would be 2 : 1. Relatively large numbers of offspring might be needed to distinguish between a 3 : 1 and a 2 : 1 ratio.

allele (CC^{cr}) of the basic color gene. The gene action of C^{cr} dilutes red pigment to yellow. Cremellos are homozygous for the cream allele ($C^{cr}C^{cr}$). All pigment (both black and red) is diluted to a very pale cream.

In Chapter 10 on parentage testing, we will see examples of blood group factors and blood protein types that are inherited as **codominant** traits. For such traits, allelic variants do not demonstrate dominant and recessive relationships, but are always coexpressed. Codominant inheritance of alleles can be used as an efficient and powerful tool for parentage testing.

Expected ratios, statistical tests and alternative models

Learning to predict and recognize simple 1:1, 1:2:1, 3:1 and 2:1 trait ratios among offspring is extremely useful for building genetic models of trait transmission. These ratios may be complicated, as we have seen, by the interactions of more than one gene and may not be obvious unless large numbers of offspring are available. Statistical testing (**Chi-square test**) may be needed to determine whether the observed ratios match the expected ratios. (Consult a basic genetics or statistics text for information on how to apply these tests.) It is obviously much easier and faster for an individual dog breeder to obtain statistically significant data working with litters of puppies than for a horse breeder working with a single birth per mare per year!

If the data obtained do not match expected ratios, then alternative proposals need to be considered, such as:

- The trait is genetic but the hypothesized transmission mechanism is incorrect (e.g. more than one gene may be involved).
- The trait is produced by environmental influences, not genes.
- The gene shows **reduced penetrance** (phenotype is modified by environment or other gene combinations, so that the effect of the mutant gene is not easily recognized).

- One genotypic class is lethal, so the expected ratios need to be adjusted accordingly.
- Several of these mechanisms simultaneously influence the trait expression.

Two genes and more

Since the genetic makeup of every horse consists of thousands of genes, a horse breeder will want an understanding of the complex results expected from the interaction of products of more than one gene. The now familiar Punnett square provides a model to predict the outcome when two independently inherited traits are simultaneously considered, such as sex and black/red color (Figure 6).

The four different genetic types of sperm can combine with the two types of ova to generate eight genotypic classes of offspring (four phenotypes). Due to chance any particular mating of this type may never produce any red females or males, while others may have two or three of each. When data from many matings are combined and evaluated by statistical tests, the hypothesized outcomes of the random events shown in the Punnett square will be validated.

Genetic contribution from mares	Genetic contribution from *Ee* stallions			
	XE	*Xe*	*YE*	*Ye*
XE	*XX EE* black female	*XX Ee* black female	*XY EE* black male	*XY Ee* black male
Xe	*XX Ee* black female	*XX ee* red female	*XY Ee* black male	*XY ee* red male

Figure 6: Punnett square diagram showing independent assortment of two traits. X and Y are the sex chromosomes. *E* and *e* are allelic differences producing the black/red coat color difference. Each trait shows a 3 : 1 ratio (black to red or female to male). Considered together, the phenotypic ratios are 3 : 3 : 1 : 1.

Dihybrid cross

Moving along to a slightly more complicated situation, consider a mating between dun tobianos, each heterozygous for the genes controlling the dun and tobiano patterns. Geneticists label a cross between double heterozygotes a **dihybrid cross**. Dun color is a dominantly inherited trait that modifies the expression of extension to produce horses of diluted coat color and a distinctive pattern of striping on legs and along the back. Tobiano is a white spotting pattern inherited as dominant trait. The phenotypic ratios among the offspring of such a cross between heterozygotes for two genes will be 9:3:3:1 (Figure 7).

Genetic contribution from mares	Genetic contribution from stallion			
	D TO	*D to*	*d TO*	*d to*
D TO	*DD TOTO* dun tobiano	*DD TOto* dun tobiano	*Dd TOTO* dun tobiano	*Dd TOto* dun tobiano
D to	*DD TOto* dun tobiano	*DD toto* dun tobiano	*Dd TOto* dun tobiano	*Dd toto* dun solid
d TO	*Dd TOTO* dun tobiano	*Dd TOto* dun tobiano	*dd TOTO* tobiano	*dd TOto* tobiano
d to	*Dd TOto* dun tobiano	*Dd toto* dun solid	*dd TOto* tobiano	*dd toto* solid

Figure 7: Complex interaction between two genes is demonstrated in a dihybrid cross between two heterozygous dun tobiano horses. The dun to not-dun ratio is 3:1; tobiano to solid is 3:1. Four phenotypes in a 9:3:3:1 ratio are found among offspring.

This chart does not show the interaction of these two genes with a third important color gene—extension. Try your understanding of gene interaction by drawing that Punnett square with 64 possible genotypes. To help you know when you have the right scheme, the predicted number of classes is eight, with phenotypic ratios of 27:9:9:9:3:3:3:1. (Hint: the most frequent class has at least one copy of the dominant allele for each gene and the least frequent class is homozygous for the recessive alleles for each gene.) Even more complicated examples could be constructed, but the point to be taken is that a basic understanding of the allelic actions of genes allows model building to predict outcomes of complex interaction.

Epistasis and hypostasis

Occasionally the action of alleles of one gene may cover up the actions of another. The masking of allelic effects due to interaction with alleles of another gene is properly called **epistasis**, not dominance. The gene that is masked by epistasis is said to be **hypostatic**. *Dominant and recessive are terms that refer to interactions of alleles of a single gene.* In my experience, misunderstanding and misuse of the term dominance is a fundamental source of confusion for horse breeders when predicting the coat color outcomes of matings.

To show the correct use of the terms and concepts of dominance and epistasis, consider a dihybrid cross involving gray and extension. In the presence of the dominant allele *G* the horse is gray, epistatically suppressing the expression of any extension locus alleles. In other words, *G* is dominant to *g* (not gray); it is epistatic (*not* dominant) to *E* and *e* (black/red color). *E* and *e* are hypostatic (*not* recessive) to *G*. A Punnett square provides the model for the epistatic interactions of allelic combinations at the two loci (Figure 8).

The complex interaction of genes may produce phenotypic ratios that seem at first glance to be unrelated to those expected by the simple random assortment of genes on chromosomes. In this example three colors of foals could be produced by double heterozygous gray parents: gray, black and red in an expected proportion of 12 : 3 : 1. Compare Figure 7 and Figure 8 to see how epistasis can affect number of classes and ratios expected among offspring.

Genetic contribution from mares	Genetic contribution from stallion			
	G E	*G e*	*g E*	*g e*
G E	*GG EE* gray	*GG Ee* gray	*Gg EE* gray	*Gg Ee* gray
G e	*GG Ee* gray	*GG ee* gray	*Gg Ee* gray	*Gg ee* gray
g E	*Gg EE* gray	*Gg Ee* gray	*gg EE* black	*gg Ee* black
g e	*Gg Ee* gray	*Gg ee* gray	*gg Ee* black	*gg ee* red

Figure 8: Complex interaction between different genes is demonstrated in a cross between two heterozygous gray horses. G is epistatic to extension so that whenever a G is present it will be expressed, regardless of the genotype of extension. The gray to not-gray ratio is 3 : 1. Also shown among the not-grays is the now familiar example of the 3 : 1 ratio of black and red. Taken together the colors gray, black and red occur in the ratio 12 : 3 : 1.

An alternative to Punnett squares

An experienced geneticist probably would not draw Punnett squares to predict outcomes of matings. It is much easier to multiply the expected fractional proportions for phenotypic classes. In a simple cross between heterozygotes (*Aa*), $3/4$ of the offspring show the dominant phenotype (*AA* or *Aa*) and $\frac{1}{4}$ the recessive phenotype (*aa*).

In a dihybrid cross:

- $\frac{9}{16}$ ($3/4 \times 3/4$) are expected to show the double dominant phenotype—genotypes *AABB*, *AaBB*, *AABb* and *AaBb*.
- $\frac{3}{16}$ ($3/4 \times \frac{1}{4}$) show the phenotype of one dominant and one recessive gene—genotypes *AAbb* and *Aabb*.

- $\frac{3}{16}$ ($\frac{3}{4} \times \frac{1}{4}$) show the other dominant phenotype with the other recessive phenotype—genotypes *aaBB* and *aaBb*.
- $\frac{1}{16}$ ($\frac{1}{4} \times \frac{1}{4}$) show the double recessive phenotype—genotype *aabb*.

Linkage

Although the independent inheritance of genetic traits has been emphasized, some genes will tend to be inherited together. The august monk Mendel did not envision the relationship we know to exist between genes and chromosomes. However, to understand genetics more completely it is necessary to expand his otherwise elegant theories to include gene linkage.

Any given gene has a particular chromosomal assignment and a place on that chromosome called a **locus** (plural: **loci**). Occasionally traits of interest are on the same chromosome (**linked genes**) and tend to be inherited together more often than they are split apart. Linked genes can be separated from each other as part of the normal process of chromosome recombination uniquely occurring during meiosis. At present, knowledge of the linkage groups distributed on the 32 pairs of horse chromosomes is meager (see Chapter 17, The horse gene map), but gene mapping efforts bode well for learning much more about specific horse gene linkages during the next decade.

A special case of linkage relates to genes on the X chromosome. Remember males have only one X chromosome, so X-chromosome genes in males are always **hemizygous**, while they may be either homozygous or heterozygous in females. Males may inherit from their mothers an X chromosome with a rare defective gene whose expression in females is usually masked by the normal gene on the other X chromosome.

The inheritance of **X-linked recessive** disease-causing genes has an expression pattern that reflects the transmission of sex chromosomes. X-linked genes are often called **sex linked**, to contrast them with autosomal genes which are located on any of the other 31 pairs of chromosomes.

Typically, hemizygous males inherit their defective gene from mothers who are heterozygotes (**carriers**) for the abnormal gene. Half the sons of a carrier mother and a normal father will be affected, but none of the daughters, although half the daughters will be carriers like their mother. Sons of affected fathers never receive the defective gene from their fathers. Daughters of affected fathers are usually free of the disease, but they are expected to transmit the problem to half of their sons. X-linked inheritance can be illustrated with a Punnett square where X* is used to symbolize the X chromosome with the disease-associated gene (Figure 9).

A useful gene linkage group in the horse, not on the X chromosome, includes a coat color complex of extension (*E*) and two dominantly inherited pattern genes, roan (*RN*) and tobiano (*TO*). An observant owner may discover that her black stallion heterozygous for tobiano spotting sires chestnut foals, but almost never are they spotted. This observation can be nicely explained

Genetic contribution from mares	Genetic contribution from stallions	
	X	*Y*
*X**	*XX** carrier female	*X*Y* affected male
X	*XX* normal female	*XY* normal male

Figure 9: Inheritance of X-linked recessive genetic defect (symbolized X*). Only males will be affected, by a gene transmitted from mother to son.

by linkage of the genes for tobiano and color. For the stallion of this example, one of his chromosomes has *TO* and *E* (tobiano with black); the alternate chromosome has *to* and *e* (solid color with red). For other stallions the linkage phase of tobiano may be *TO* with *e* (tobiano with red).

Understanding the concept of linkage may be useful for breeders. One of the ways for an undesirable trait to become widespread in a breed is through close linkage with a highly prized trait in popular bloodlines (**founder effect**). A foal with the valued gene without the linked defective trait may be quite rare if the genes are closely situated on the same chromosome.

Polygenic (multiple gene) traits

Most traits are influenced by more than one gene. Coat color provides a good example of how useful it can be to break down the phenotype into components that help us recognize the presence of alleles of the various genes involved. Some traits cannot be as easily broken down into the gene components from phenotype alone. Conformation traits are examples of multiple gene interactions whose components are unknown both as to number of loci involved and number of alleles at those loci.

Meat, milk, fiber and eggs are production traits of cattle, sheep and poultry that can be measured (**quantitative traits**). The phenotype can be produced by additive effects of genes at several loci and also typically influenced by environmental components. Computer-based mathematical analysis provides genetic models for quantitative traits to identify elite breeding stock based on their own production records as well as those of their offspring and close relatives. Current active research in cattle is also having some success to define **QTLs** (quantitative trait loci) for milk production. The studies look for correlated presence of high levels of milk production with specific DNA gene markers. Milk genes are good candidates for QTL studies since large families, production records and a comprehensive map of DNA markers are readily available for cattle.

Genetic analysis of polygenic traits has not yet proven to be as useful for horse breeding as it is in livestock production. A good DNA genetic marker map for horses is a current collaborative research project of several laboratories worldwide and may eventually help with the genetic analysis of equine performance traits.

Determination of inheritance patterns

To apply genetics to a breeding program requires understanding of how traits are inherited before designing the selection scheme. A **familial trait** is one that has been observed to occur in families but is not necessarily inherited. It might be caused by shared environmental factors (diet, infection, or toxins). How can one determine whether a trait is inherited and whether it is dominant, recessive, sex-linked, autosomal, or polygenic? It is important to wait for results of breeding trials to prevent making decisions based on tentative but erroneous conclusions. Several years of research may be necessary to provide definitive answers. Components to the research include:

- Searching the scientific literature for a well-described candidate gene that has similar clinical and pathological characteristics in other mammalian species.
- Collecting pedigrees of horses expressing the trait to search for relationships.
- Collecting tissue samples (blood or hair roots) to store for possible future genetic marker research (especially to look for DNA markers).
- Collecting offspring data from breeders to suggest the pattern of trait transmission.
- Designing matings and performing crosses to test genetic hypotheses, an essential, major money-consuming aspect of any research proposal.

The candidate gene approach can be extremely useful, particularly for disease conditions, and may readily yield a diagnostic test for carriers if a defective protein or DNA sequence has already been identified in another species. Pedigree collection is useful, but since purebred horses are likely to be related to each other and to be produced from matings between stock at least distantly related, spurious conclusions may be reached, particularly by people unfamiliar with the breed structure. On the other hand, a deleterious gene may be a chance hitch-hiker in the genomes of horses highly promoted or fashionable. The tendency of breeders repeatedly to use selected sire lines can raise the gene frequency of a deleterious recessive to such a level that matings between carriers are common enough to produce a detectable problem.

An **autosomal dominant** trait will show direct transmission from parent to offspring (except in the unlikely event the trait has just appeared as a new mutation). About 50% of offspring of both sexes will have the trait. If the trait is a disease problem, normal offspring of affected parents when mated to a normal mare or stallion produce only normal foals. The trait will generally be rare and confined to a single breed. It will be frequent in the family in which it occurs, even if the family shows little inbreeding.

An **autosomal recessive** trait initially will probably be seen in both sexes as a result of matings between horses without the trait (usually related). It will generally be confined to one breed or to closely related breeds. This kind of genetic trait can appear to skip generations. When animals with the trait are mated together, all the offspring will show the trait. The trait will be rare within the breed, but more frequently seen in inbred families.

An **X-linked recessive** trait generally is obvious from an apparent association with males, but is not a trait necessarily associated with sexual characteristics.

A **polygenic** trait will be more frequent among related horses than among horses in general, has no obvious inheritance pattern and usually is not confined to a single breed. A polygenic trait may show a continuous range of variation, typically graded from normal to slightly affected to extremely affected. Examples of such traits in humans include diabetes and heart disease and for horses would include conformation traits. An alternative pattern for a polygenic trait is to show a threshold effect. The trait is either present or not but is not inherited in the fashion of a single Mendelian gene. For both polygenic trait patterns usually a few major genes are involved whose actions or effects can be modified by other genes and by environmental components such as nutrition.

CHAPTER 2
Black, bay and chestnut

Probably the ancestral color of the horse was a black-based pattern that provided camouflage protection against predators. The coat might have resembled that of Przewalski's horse, the surviving wild species most closely related to domestic horses, typically described as dun. Whatever the ancestral color, under domestication the horse has clearly evolved into an animal with a wide range of colors. The distinctions among these colors are controlled by genetic differences that occurred as mutations and have been selected for during the course of domestication.

In mammals **melanin** is the most important pigment of coat color. It occurs as pigment granules in the hair, skin, iris and some internal tissues in two related forms: **eumelanin** (black or brown) and **pheomelanin** (red or yellow) (Searle 1968). The biochemistry of pigment production in the horse is homologous to that of other species (Woolf & Swafford 1988). Genes produce coat color variation by altering the switch between eumelanin and pheomelanin production in pigment cells (**melanocytes**); or the presence, shape, number or arrangement of pigment granules.

Color names for most phenotypes are easy to designate by visual inspection. Owners and breed registries record colors as one means of identifying individuals. Colors help suggest a particular animal's genes for pigment synthesis and distribution, but remember the limitations of word descriptions:

- At present no laboratory test is available to verify or assign any color gene of the horse. Words provide an approximate description of the genes involved and a working model to predict the color outcome of different matings. Two horses with the same word description may not have the same genotype.
- Most breeds have a rather restricted set of recognized color names compared to the variation of hues that exists. Although frustrating at times, this may be an appropriate policy since much of the variation in shade and intensity within colors is not yet defined by adequate genetic models.
- Two breeds may use different names for colors that appear to be the same or the same name for colors that are genetically different.

Traditionally speaking, the basic colors of horses are black, bay and chestnut (red), if for no other reason than that the majority of horses can be described by these terms. Near the turn of the nineteenth century, Hurst (1906) showed that Thoroughbred studbook color records could be explained using genes inherited according to Mendelian principles. The genes extension (*E*) and agouti (*A*) are currently assigned to account for inheritance of the differences among the basic colors.

Extension

In the horse extension is credited with producing the difference between black pigmented horses (blacks, browns and bays; also buckskins, duns and grullas) in comparison with the reds (chestnuts and sorrels; also palominos and red duns). Formerly the difference was attributed to a brown (*B*) gene, but the extension locus terminology is now preferred in light of homologies between pigment genetics of horses and other mammals. (Whether horses have genetic variation at a gene homologous to brown in other mammals is presently undetermined.)

Trait inheritance and gene symbol

Two alleles of extension are assigned to account for the black/red color variation in horses. The alleles extend (*E*) or diminish (*e*) the amount of eumelanin (black) in the coat with the opposite effect on the extent of pheomelanin (red). For an understanding of the extension gene, the gene action to consider is whether black pigment is found in hair and skin (*EE* or *Ee*) or only in the skin (*ee*). A major feature of black hair pigment in horses is that it may be distributed either uniformly or in a pattern (black hair in mane and tail and on legs; but reduced or absent on the body). Black pattern characteristics are due to a second gene (agouti), to be discussed in the next section.

Black, brown or bay:	*EE* or *Ee*
Chestnut (red):	*ee*

The phenotype of *ee* horses ranges from deep chestnut to light sorrel, but the genetics of these variations is presently undefined. Chestnuts in particular may show seasonal variation, with the spring coat emerging from shed winter hair being markedly darker than that of late summer and winter. Some reds have a conspicuously light (flaxen) mane and tail. This effect is probably due to actions of yet another gene that appears to affect bays as well. Many red horses have small, irregular dark hair patches that look like smudges on the red coat ("Bend Or spots") (Figure 10). Eumelanin production in hair appears to break through in these *ee* horses, but not to an extent that they could be confused with *E*–'s.

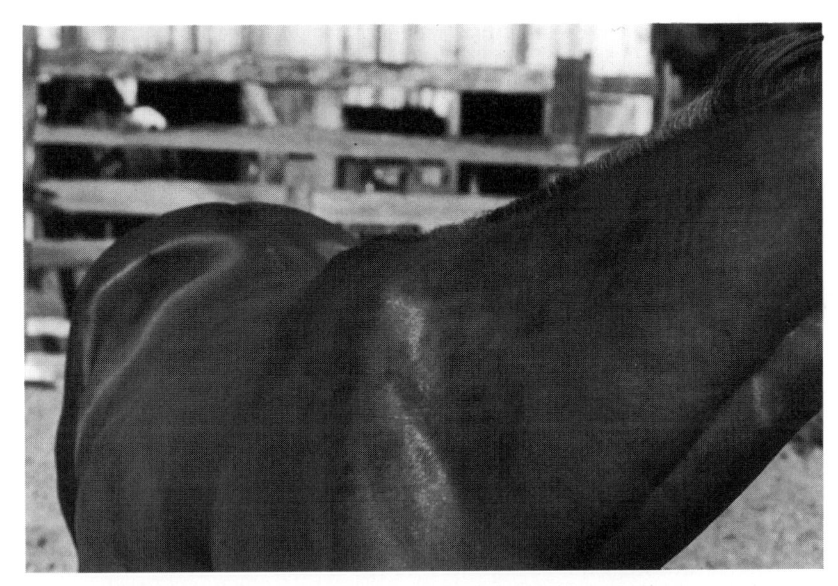

Figure 10: Blotchy dark spots on a chestnut (*ee*) are sometimes called Bend Or (Ben d'Or) spots after a Thoroughbred stallion with these markings.

The presence of black pigment is inherited as a trait dominant to its absence, so matings between two red (*ee*) horses should not produce any black/brown/bay offspring. This "chestnut rule" has been repeatedly verified by parentage exclusion in exceptional cases (Trommershausen-Smith *et al.* 1976a).

As with the reds, great variation is found among the black/brown/bay category, forming a graded series subject to much discussion and even arguments concerning the designation of selected horses as examples of the "true" colors (e.g. black and brown). Such discussions are likely to be inconclusive until an objective test that identifies specific alleles (by DNA sequence?) is available.

Extension is a good candidate, as in mice, to have several alleles that affect black hair distribution. A third allele in horses (E^D) has been proposed to account for "dominant black" that may occur in some breeds, but it has not been studied extensively.

Some breeders are keen to locate stallions that sire only black-pigmented foals. To be homozygous for black pigment (*EE*), a first requirement is that both parents must have black pigment (*E*–). Remember that black-pigmented horses (*E*–) could be black, bay, brown, buckskin or grulla. Among such matings at least 25% of the offspring will be *EE*. A test cross to red mates will identify the homozygotes. The statistical assurance of homozygosity is related to the number of offspring available (Table 1).

Not all foals of a black stallion homozygous for black pigment (*EE*) will necessarily be of the color called black. When the dam contributes the appropriate pattern genes, the foals could be black, bay, brown, buckskin or grulla.

Number of black-pigmented offspring, no reds, from matings of black to red	Prediction certainty for black-pigmented parent to be homozygous *EE*
5 black-pigmented foals	97%
7 black-pigmented foals	99%
10 black-pigmented foals	99.9%

Table 1: Statistical assurance of extension gene homozygosity prediction from test cross breeding (black-pigmented × red). Accuracy of prediction for black-pigment homozygosity (*EE*) is related to number of offspring.

Any chestnut, palomino or red dun foal, regardless of the color of the dam, would prove the black stallion must be *Ee*.

Gene linkage

Family studies assigned extension to the second group of linked genes discovered in the horse (linkage group II, LG II) (Andersson & Sandberg 1982). The chromosomal assignment of LG II is not currently known, but this interesting group includes other coat pattern genes roan (*RN*) and tobiano (*TO*).

Gene homology

A eumelanin/pheomelanin switch is seen in the hair colors of many mammalian species and is a prominent source of color variation among domestic animals. Labrador Retriever dogs may be black or yellow (or chocolate, but that is a gene story we will not be discussing in the horse context). Holsteins and Angus are predominantly black-pigmented cattle breeds, but red animals occur and some owners specifically breed for the red color.

From studies in mice it is known that extension codes for a protein that is part of the membrane of melanocytes, the cells that make pigment. This protein binds a hormone that stimulates the cell to make eumelanin. The extension gene has thus been defined as the DNA sequence coding for the melanocyte-stimulating hormone receptor. Recessive mutants of extension in the mouse fail to bind the melanocyte-stimulating hormone, so only pheomelanin, not eumelanin, is made (Robbins *et al.* 1993). Other mutants bind the hormone so tightly that eumelanin is always made. Perhaps some shades of bay and brown in horses will eventually be defined genetically as variants at this locus. The DNA sequence of the mouse gene has been determined and this information may soon be applied to determine extension genotypes in horses.

Breeds

Most breeds include both red- and black-pigmented variants, but a few breeds have been selected to have little or no variation for this gene. Friesians and Cleveland Bays have only the dominant allele—rare recessives could be present but extremely difficult to detect—and Suffolks and Haflingers have only the recessive.

Agouti

Agouti controls the distribution pattern of eumelanin, so its actions are obvious only in the presence of *E* (an example of epistatic gene interaction).

Trait inheritance and gene symbol

The dominant allele *A* causes the distribution of eumelanin in hair to be restricted to a "points" pattern (e.g. mane, tail, ear rims, lower legs) (Figure 11). The recessive allele *a* does not restrict the distribution of black hair and when homozygous in the presence of *E* produces a uniformly black horse.

Bay or brown:	*EEAA*, *EeAA*, *EEAa* or *EeAa*
Black:	*EEaa* or *Eeaa*
Chestnut (red):	*eeAA*, *eeAa* or *eeaa*

Breeders specializing in black horses will want to understand the special combination of two genes required to obtain the color. Predicting whether given matings will produce black is complicated by the uncertainty about

Figure 11: Black pigment prominently displayed in a pattern mostly confined to the points (mane, tail, lower legs and ear rims), as in this bay Arabian stallion, is attributed to actions of the dominant allele *A* of the agouti gene.

which alleles of agouti are present in red (*ee*) horses. In many breeds, the *a* allele is rare (black horses are infrequent), so most bays and reds can be assumed to be *AA*. Any red or bay horse that sires or produces a black offspring must be carrying *a*. If a red horse has two black (*Eeaa*) parents, obviously the red horse must be *eeaa*.

Black coat color varies from blue-black to sun-fading black, but the genetic differences among the variations are currently undefined. Other alleles hypothesized at agouti in horses (A^+, A^t) have been proposed to define the inheritance of wild pattern (as in Przewalski's horse) and brown, but they have not been studied extensively. Alternative alleles of agouti might be responsible for some of the variation in pigment distribution of color shades of bays, browns and blacks.

Gene linkage for agouti is currently undetermined. Even though agouti and extension both control eumelanin production, they do not appear to be genetically linked.

Gene homology

The gene derives its name from a South American rodent with black-banded hairs. In mice the gene action is known to produce a molecule that regulates the production of eumelanin versus pheomelanin synthesis, either during hair growth phases (black-banded yellow hairs) or spatially (black hairs on the back, but not on the abdomen). It has been assumed that the uniform distribution of black hair in black horses compared with the restricted black (point) pattern of bays is due to a gene homologous to agouti in rodents, but homology at the molecular level has not been established. Extension and agouti are not genetically linked in the mouse.

Breeds

The uniformly bay Cleveland Bay breed has only the *A* allele, and black Friesians probably have only *a*. Most light horse breeds appear to have a higher frequency of *A* than *a*, but for pony and draft breeds the frequency relationship may be reversed.

CHAPTER 3
Color diluting genes

Coat color dilution in horses is due to the actions of at least three genes. These genes are best understood individually, but it is possible for a horse to have the variant alleles of more than one dilution gene. Most phenotypes for these genes in combination with the basic coat colors are well known. Some infrequent combinations could use keen sleuths to find the examples, so we can learn to recognize the effects of gene interactions.

Cream (palomino and buckskin)

Probably the most widely recognized of the color dilution genes is the one that produces the golden body color seen in palominos (Figure 12) and buckskins (Figure 13). A palomino horse has a white (flaxen) mane and tail and a buckskin has black mane, tail and legs. The palomino is a genetically modified (diluted) sorrel or chestnut and the buckskin a modified bay. Sometimes the mane and tail of the palomino or the body color of the buckskin have dark hairs intermixed, reflecting similar admixtures found in the undiluted dark bay or chestnut counterparts. The skin and eyes of palominos and buckskins are dark, although they may be lighter than those of non-diluted colors. Cremellos and perlinos also belong to the palomino/buckskin color family. Cremellos have pink skin, blue eyes and ivory hair. Perlinos have the same features except the mane and tail are darker than the body.

Palominos can be very light cream when born. A darker golden color becomes obvious after the foal coat is shed. Buckskins may also be light when born, even to a degree that the black points may not be obvious until the foal is some weeks old. These colors may also vary seasonally, being lighter in winter coat.

Trait inheritance and gene symbol

Odriozola (1951) proposed that an allele of the color (albino) (*C*) locus, which controls tyrosinase action and melanin production in other mammals, is responsible for the golden color of palominos and buckskins, and the extreme dilution to cremello and perlino. The C^{cr} (cream) allele shows incomplete dominance, diluting pheomelanin (red) to yellow when heterozygous

Figure 12: Palomino pony (*eeCC*^{cr}) showing golden body color and white (flaxen) mane and tail.

but having little or no effect on eumelanin (black). Both eumelanin and pheomelanin are diluted to pale ivory when the dilution allele is homozygous ($C^{cr}C^{cr}$) (Adalsteinsson 1974).

Not dilute:	CC
Palomino/buckskin:	CC^{cr}
Cremello/perlino:	$C^{cr}C^{cr}$

Combining the dilution gene symbols with those for the coat color genes involved with the basic production of color, palominos are diluted reds (CC^{cr} ee) and buckskins are diluted bays (A– CC^{cr} E–). Blacks can carry the dilution gene without expressing it (aa CC^{cr} E–), since they do not have red pigment to show the single dose effects of the dilution gene.

Why the palomino has a white mane and tail, not gold like the body color, is not easily explained, but could be related to differences in gene action between melanocytes (pigment producing cells) of permanent hair (mane and tail) compared with seasonally shed hair.

It has been proposed that cremellos are homozygous dilute reds and the dark-maned perlinos are homozygous dilute bays or blacks, but that explanation does not fit all examples. Homozygous dilute bays ("perlinos") can be a uniform ivory color, indistinguishable from their homozygous dilute red ("cremello") counterparts. More cases need to be studied to understand the genotypic and phenotypic differences between cremellos and perlinos.

Although the palomino/buckskin color dilution factor usually behaves as an incomplete dominant, some apparent exceptions are known. Black horses (without pheomelanin) may carry the C^{cr} allele showing only subtle phenotypic dilution effects, such as yellow-brown eyes and a dulling of the black body color. Such blacks may unexpectedly sire or produce dilute foals from non-dilute mates, but the source of the allele should nonetheless be traceable through the pedigree behind the black parent.

Other very rare exceptions are found, such as a cremello from a bay × palomino mating, that do not have a completely satisfactory explanation. Some breeds of horses may have recessive dilution alleles of the *C* gene, similar to the recessive albino seen in many other mammals. Candidates for other allelic variants of the palomino/buckskin family include very pale palominos, sometimes called isabellas, and pink-skinned amber horses, sometimes called yellow duns. Alternatively, these light shades may be due to other genes that modify pigment intensity.

Novice breeders captivated by the beauty of the palomino and buckskin colors are discouraged to learn that neither will breed true since the desirable colors are produced by heterozygosity for a dilution gene. If a breeder attempts to duplicate the color of a favorite palomino, say by breeding that palomino to another palomino, the predicted colors and their frequencies among the offspring will be 50% palomino, 25% red and 25% cremello. Breeders of Quarter Horses and Morgans should be aware that a cremello (or perlino) will not be accepted for registration. For these breeds, the better mating choice would be palomino × red—the expected proportion of palominos is the same as from a palomino × palomino mating (50%) but no cremellos would be anticipated.

In breeds that allow registration of cremello and perlino colors such as Icelandic, Miniature, Peruvian Paso or Paso Fino, a homozygous dilute horse can be an important component for a breeding program specializing in palominos and buckskins. The cremello or perlino will contribute a color dilution gene to all its offspring (Figure 13).

The gene linkage group for *C* in the horse is currently undetermined.

Figure 13: A homozygous dilute mare ($C^{cr}C^{cr}$) (right) with her buckskin daughter (*A–EeCC*^{cr}) (left) by a chesnut (*eeCC*) Arabian stallion.

Gene homology

In mice and humans, albinism historically was attributed to genetic variation at the color (*C*) gene. Progress in molecular biology has now more specifically defined color dilution genes. The tyrosinase (*TYR*) and pink-eye (*P*) genes have several alleles with graded effects, from very slight color dilution to extreme (albino). In other mammals, the *C* gene terminology has been retained for the moment but probably will be replaced by *TYR* or *P* when molecular-level homology is demonstrated.

Siamese, Himalayan and Burmese cat colors are produced by *C*-series alleles (Searle 1968). These alleles show temperature-dependent effects. The color dilution is seen most prominently in hair on the warmer parts of the body, in contrast to the colder extremities. Dilution colors in cats are produced by recessive alleles.

If the palomino/buckskin dilution gene is at the tyrosinase locus, then hair bulbs may show a deficiency in tyrosinase activity, as demonstrated for some types of albino in humans. We made a preliminary study to test the assignment of horse color dilution using hair from a cremello horse. No tyrosinase activity was seen in hair bulbs from this horse incubated in L-dopa (a chemical precursor to melanin), while a palomino and normally pigmented horses showed blackening of hair bulbs under the same conditions.

Breeds

The palomino/buckskin dilution occurs in a variety of breeds. It is typically associated with ponies and stock horse breeds but also occurs in Paso Finos, Peruvian Pasos, American Saddlebreds, Morgans and Tennessee Walking Horses. Palominos and buckskins occur among US Thoroughbreds, but they are very rare. This dilution gene is probably absent from Arabians. Although Arabian iridescent light chestnuts with extremely flaxen manes and tails may be registered with palomino societies, the buckskin and cremello counterpart colors to palomino are not seen in this breed.

Defects associated with cremello/perlino

The blue eyes of cremellos and perlinos are sensitive to the sun and owners report these horses actively seek protective shade in summer (photophobia). Pink-skinned horses are more subject to neoplastic conditions such as squamous cell and basal cell carcinomas about the eyes (Knottenbelt & Pascoe 1994).

Dun

In an otherwise red horse, dun produces a pinkish-red horse with darker red points and the complex pattern of dorsal stripe, shoulder stripe and leg bars. This color is known as red dun or claybank dun. In an otherwise bay animal, dun produces a more or less yellow-red animal with black points and the pattern of dorsal stripe, shoulder stripe and leg bars. This color is known as buckskin dun or sometimes simply dun (Figure 14). An otherwise black animal with the dun gene is a mouse-gray color with black points and the pattern of dorsal stripe, shoulder stripe and leg bars. This color is known as mouse dun or grulla. Sponenberg and Beaver (1983) provide color photographs and a discussion of the wonderful variety of names attached to this series of colors.

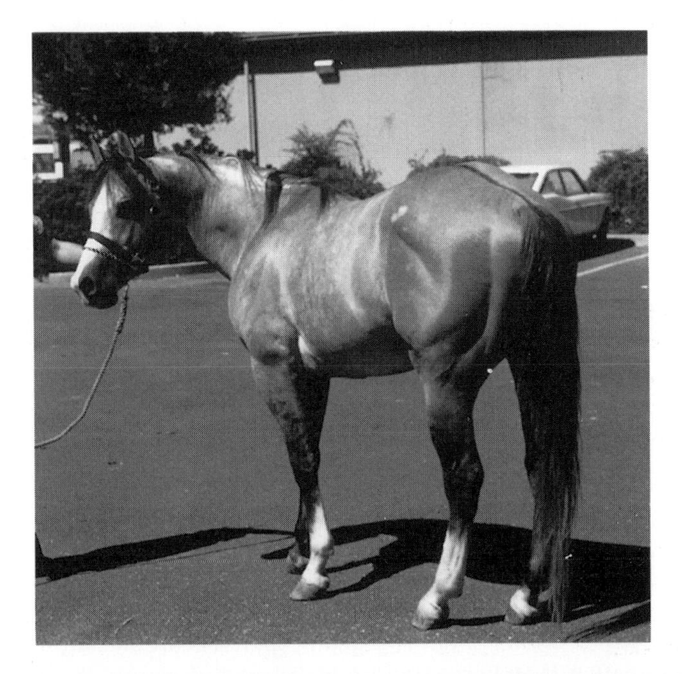

Figure 14: The colour of this buckskin dun (*A–E–D–*) Quarter Horse is similar to the buckskin in Figure 13, but the dun pattern characteristics (dorsal stripe, shoulder stripes and leg bars) are the clues that the diluted color of this horse is produced by the dun gene.

Dun can also be used as a catchall category to include horses of the palomino/buckskin series, the line-backed dun series and probably some currently undefined hues. In this chapter the term dun is used in the narrow sense, only for diluted horses with the line-backed dun pattern characteristics.

Trait inheritance and gene symbol

The dominantly inherited dun trait dilutes both eumelanin and pheomelanin of body hair, but does not dilute either pigment in hair on the points. Red body color is diluted to pale red (claybank or red dun) or yellow red (buckskin dun); black body hair is diluted to mouse-gray (grulla) (Van Vleck & Davitt 1977). In addition to pigment dilution, *D* is characterized as producing a coat pattern which includes dark head, dark points, dorsal stripe (list), shoulder

Figure 15: Web-like marking on shoulder of a dun-colored Przewalski's horse.

stripes, and leg bars. These features may show variation between animals (number of stripes, width of stripes), but the genes that modify the pattern are presently unknown. Among some Przewalski's horses and Mongolian ponies, the dun markings on the shoulder and upper leg may be in the form of a distinctive network or webbing pattern (Figure 15).

Homozygotes for *D* are phenotypically indistinguishable from heterozygotes and do not show the extreme color dilution effects associated with homozygotes for the palomino/buckskin gene. Gremmel (1939) showed that in the hair shafts of duns, pigment granules are heavily concentrated on one side rather than being uniformly distributed around the core.

Dun:	*DD* or *Dd*
Not dun:	*dd*

The *D* allelic effects can be confused with those of C^{cr}; however, there are several important differences. First, *D* dilutes both black and red pigment on the body but does not dilute either pigment on the points. Red body color is diluted to a pinkish-red, yellowish-red or yellow; black body color is diluted to mouse-gray. Second, in addition to pigment dilution, a characteristic of *D* is a striping pattern. Third, homozygosity for *D* does not produce extreme color dilution. This gene affects clumping of pigment granules, thus providing an optical dilution effect, in contrast to C^{cr}, which probably controls chemical alteration of pigment.

A horse may have both the C^{cr} and *D* dilution alleles. A red horse with both dilution genes looks like a palomino with dun markings. Heterozygosity for dilution genes at both loci does not result in extreme color dilution.

More or less faint dun markings can be found on some otherwise "ordinary" undiluted chestnuts and bays and may be prominent on young grays. These dun markings without color dilution are probably the effects of another gene, but at times the effect can be confused with *D*.

The gene linkage group for dun is currently undetermined.

Gene homology

In other mammals (Searle 1968), *d* is the symbol for dilute, a recessively inherited color gene that effects a distinctive clumping of pigment granules producing the optical effect of color dilution. If mapping efforts are eventually successful to define linkage relationships for dun of the horse, the appropriateness of the *D* nomenclature may be confirmed or it may be changed for a more suitable gene terminology that reflects gene homology.

Breeds

In North America the *D* allele is seen in stock horses, in ponies and among feral horses. A prominent breed example of dun is the Norwegian Fjord. There are also color breed registries restricted to horses of the palomino/buckskin and dun colors. The dun trait is generally found in breeds that also have C^{cr}, so it is important to be able to distinguish the variants of these genes both alone and in combination.

Silver dapple

A third color diluting gene, silver dapple, was described by Castle and Smith (1953) and purported to have originated in the late 1800s among Shetland Ponies. This origin has been widely accepted, but the occurrence of the color in Icelandics, likely to share an historical origin with Shetlands, but certainly with no interrelationship in the last several hundred years, refutes Castle and Smith's named founder gene source. The color occurs (rarely) in horses as well as ponies, probably more often in Europe than in North America.

The gene effects are seen conspicuously on genetically black horses in which the coat color is diluted to a chocolate or black-chocolate, often with dapples, and the mane and tail are diluted to silver gray or flaxen (Figure 16). On genetically bay horses, the gene produces color dilution so that the horse is usually described as a silver-maned chestnut, but the legs retain some darker pigment. A possible color name for a bay with the silver dapple dilution is silver bay. The gene probably has little effect on chestnut (pheomelanin) coat color, beyond producing a silver (flaxen) mane and tail. This color is sometimes called silver sorrel, but it is difficult visually to distinguish from sorrel.

The addition of the word "dapple" to the name of this dilution gene is unfortunate. The name implies that dapples are always associated with this color, which is emphatically not true. The word "dapple" leads some owners

Figure 16: Dilution of black by the silver gene gives a chocolate coat with flaxen or silver gray mane and tail. In this Miniature Horse mare the dilution effect does not include dappling, commonly included in the color name.

to confuse this color with gray. In Miniature Horses many silver dapple horses are registered as dapple gray, probably in part because the color option for silver dapple was only recently provided.

Classic silver dapple foals are born a light reddish brown color with the same color mane and tail. As the foal coat is shed, the mane and tail grow in light and the body color darkens to deep chocolate. Silver dapple interacts with gray so that the birth color for foals with the gray allele (*G*) is comparable to a mature gray.

Trait inheritance and gene symbol

The trait is inherited as a dominant, but the gene action in combination with variants at other coat color genes is poorly documented (Figure 17). Castle and Smith (1953) proposed *S* for the gene symbol, reflecting the name silver given to the color in horses, but this symbol is not suitable because in other mammals it is used for spotting traits. The silver dapple phenotype shares some similarities with palomino (light mane and tail, dilution of body color), but no genetic tests of the relationship are reported. The gene may show dosage differences between homozygous and heterozygous states. Currently, *Z* is often used as the gene symbol for silver dapple, but a standard terminology has not been adopted.

Bay or black exceptions to the chestnut coat color rule (Trommershausen-Smith *et al.* 1976a) could be explained by the presence of silver dapple in one parent. That is, an *A–E–* bay with the silver dapple gene (a silver bay) could appear to be a chestnut (darker points than typically associated with chestnut, but for lack of a better category may be registered as chestnut). When the silver bay (registered as chestnut) is bred to a chestnut, offspring receiving *E* but not the dilution gene, would be—legitimately—bay or black. Due to the

Figure 17: An Icelandic horse demonstrates the combination of dun and silver on a bay genetic background. This gene combination is rarely encountered among horses known to the English-speaking world and thus no special name is available for this distinctive genotype/phenotype.

rarity of the silver dapple gene in horses, probably most exceptions to the chestnut rule will be due to incorrect parentage. Genetic marker testing to identify the true parents would be an appropriate effort for pedigreed horses.

The genetic linkage group of silver dapple is currently undetermined.

Gene homology

No homologous gene has been identified. Silver in other mammals (e.g. cats) describes a different color dilution effect, so an *Si* symbol probably is not appropriate.

Breeds

In North America, this color dilution gene is most conspicuously found in Shetland, Miniature and Icelandic breeds and perhaps (rarely) in Quarter Horses, Morgans and Peruvian Pasos.

CHAPTER 4
White, gray and roan

White

Horses popularly called white or albino are produced by the action of at least three genes. Classically, in the presence of the dominant allele at the white locus (*W*), a horse from birth will lack pigment in skin and hair. The skin is pink and the hair white, but the eyes are usually dark brown (Figure 18). Small black spots may be found in the skin, but usually are not accompanied by colored hair. Included in the everyday use of "white" are older horses produced by the action of *G* (gray) and "albinos" (cremellos and perlinos) that are homozygotes for the C^{cr} dilution gene. Those types are clearly distinguishable from *W* horses. Every effort should be made by persons interested in understanding horse genetics to avoid use of the color term white when genetically inappropriate.

Trait inheritance and gene symbol

All non-white horses are *ww*. As far as is known no horse is homozygous for *W*. Study of white horse breeding data showed that in repeated matings between white horses, both solid color (non-white) and white foals were always produced. The ratio of white to colored foals more closely approximated the 2:1 ratio anticipated for a gene with a homozygous lethal class (presumably *WW*) than the 3:1 anticipated if all offspring were equally viable (Pulos & Hutt 1969). Apparently the *WW* embryo or fetus dies early in gestation and is either resorbed or aborted without a trace. Wriedt (1924) reported sterility associated with white horse breeding in Denmark at the Fredericksborg Stud, hypothesizing a gene called the Fredericksborg lethal.

White:	*Ww*
Not white:	*ww*

Figure 18: Pink skin and dark eyes distinguish this Thoroughbred mare as white (*Ww*), not gray (*G*–) or cremello ($C^{cr}C^{cr}$) (from Bowling 1992, reprinted with permission of publisher, Elsevier).

Dark-eyed white horses may rarely occur as offspring of solid (dark) parents, for example in Thoroughbreds, Standardbreds and Arabians. At birth these horses usually have pigmented hair on and around the ears, in the mane and on the back, but the pigment may disappear with age. White horses from dark parents transmit white as a dominantly inherited trait. Their offspring may be white or dark; occasionally the whites have patches of colored hair (Figure 19), usually intermixed with white. One explanation for white horses from dark parents is that they arise from a gene mutation that can be transmitted to their offspring. The variegated pattern in some offspring may be a display of genetic reversion in some melanocytes back to the "normal" state.

Figure 19: A Thoroughbred foal by a bay stallion shows the variegated form of white color. His dam has a similar pattern. She had a bay sire and a white dam.

Such gene behavior has not been investigated in horses, but in mice can be associated with gene duplication or with exogenous sequences of viral DNA incorporated into or excised from specific sites in the organisms' DNA.

At least two kinds of white horses are found in Paints. One type is only seen in foals produced occasionally from matings between overos. With rare exceptions, these foals do not survive because of a defect in nerve cells in the intestinal tract. A second type occurs occasionally as offspring of breedings between tobianos and overos. The tobiano/overo type usually shows some pigmentation about the head, but in rare cases this pigment can be lacking and is not associated with a lethal condition. Viable white stallions from Paint breedings do sire solid dark foals (as well as tobianos and overos), so this kind of white horse also appears not to breed true for white. Most probably, the classical *W* allele does not occur in Paints.

The relationship among the genes involved in producing the various kinds of white horses is not presently known. It is tempting to speculate they may be alleles of one gene, since their actions are similar and some (all?) are involved with inviability of homozygotes.

White is epistatic to all other coat color genes, obscuring their actions so that it is impossible to determine by looking at a white horse what other coat color genes it has.

The gene linkage group for white has not been determined.

Gene homology

In mouse and pig, white color is dominantly inherited and the responsible gene is linked to *ALB*. In mice *W* has lethal or deleterious effects in homozygotes, associated with mutations in *KIT*, a gene encoding the tyrosine kinase transmembrane cellular receptor for the mast/stem cell growth factor.

Breeds

The *W* allele is rare in nearly all breeds of horses. The color is found in Tennessee Walking Horses, American Albinos and Miniatures, and rare examples that may be related to white gene mutation have occurred in Thoroughbreds, Arabians and Standardbreds.

Gray

Everyone is familiar with progressive changes of human hair color in which the hair color of youth is replaced with gray or white. Horses show a similar phenomenon of hair silvering occurring at proportionately much younger ages than in people. Studbook records made on the basis of foal colors at about six months usually accurately reflect the gray color of the mature horse.

A young horse that has the progressive graying allele can be born any color. In the popular literature the birth color of grays is often said to be black, but that is not true for all breeds. The birth color of a gray horse is a function of the allelic frequencies of the other coat color genes in the breed, particularly agouti and extension. For example, most gray Arabians are born bay or chestnut, not black.

Soon after birth a foal going gray will begin to show intermixed white hairs that proportionally increase in number with age. At intermediate stages many gray horses show a dappling pattern of light gray hair splotches surrounded by dark gray rings. The knees, hocks and fetlocks may be obviously dark gray, a character usually retained longer than dappling. At maturity, the hair coat will be a clear gray (appears as a "pure" white horse with dark skin) (Figure 20) or gray with colored flecks ("flea-bitten," "mosquito-bitten,"

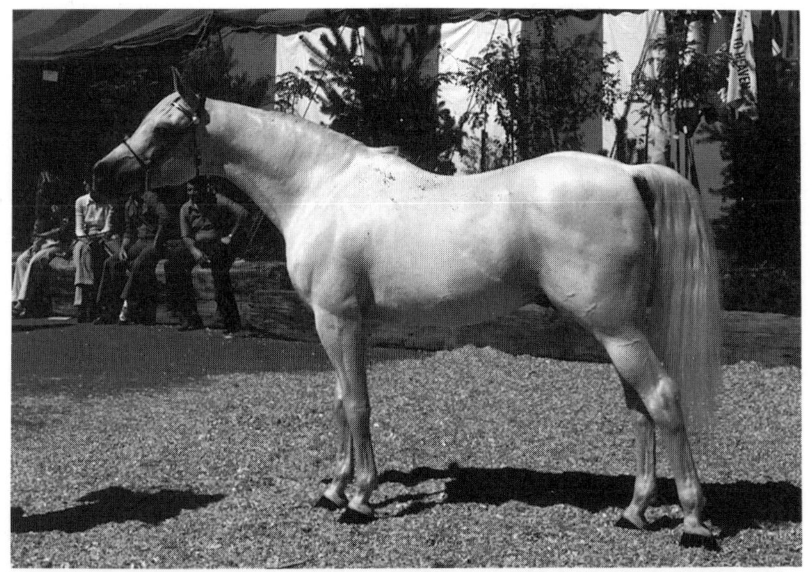

Figure 20: Mature gray (*G*–) Arabian stallion showing the black skin that clearly distinguishes this color from either a pink-skinned white (*Ww*) or a pink-skinned ivory-haired cremello ($C^{cr}C^{cr}$).

"speckled"). If flecking occurs, the small colored spots will provide clues to the base coat color produced by other genes that is obscured by the action of the gray allele. Dark pigment remains in the eyes and skin even when the hair color is completely white (except for rare examples of blotchy skin depigmentation). Occasionally, a mature gray may have colored hair patches. Large colored areas are often called "bloody shoulder marks," although they need not be confined to the shoulder area.

Care must be taken during the registration process to distinguish those white markings on a gray horse that have underlying pink skin from those that do not. Particularly about the face, white markings may not be accompanied by pink skin and in a mature gray horse such markings will no longer be visible. Tobiano, overo and appaloosa patterns may not be readily visible as a coat pattern in a mature gray horse, but can be seen as pink skin patterns, particularly when the hair is wet.

Trait inheritance and gene symbol

In horses gray color is inherited as a dominant trait. A gray horse will be either *GG* or *Gg*. It is not possible to tell by looking at the horse whether it is homozygous for *G*. A horse without the gray gene is symbolized as *gg*.

Gray:	*GG* or *Gg*
Not gray:	*gg*

Earliest indications of the *G* allele's presence can be seen by careful study of the head of a young foal, particularly around the eyes. Later, the horse will have a mixture of white and dark hairs throughout the body, a stage that can be confused with roan. Foals with the *G* allele are born much darker than their not-gray counterparts. A chestnut foal is typically born a light fawn color, particularly evident on the legs, but sheds out a darker shade. Newborn chestnut foals that will later turn gray are much darker than chestnuts that will not change to gray. The coat darkening is also seen in bay horses with *G*. Bay foals also typically have fawn-colored hair on their legs, shedding out to black. In bay foals that are going to be gray, the legs are black at birth. Black horses with *G* are born a shiny black, rather than the more typical mouse-gray foal color of blacks.

Some Arabian chestnuts going gray proceed through a stage known as "rose gray." "Dun-like" markings, often faint in other colors, may be quite prominent in gray foals. The speed of the graying process may be under genetic control. It may also be related to the base coat color. Darker colors in the presence of *G* "gray out" more slowly than do the lighter ones.

Since gray color is produced by the action of a dominant gene, at least one parent of a gray horse must be gray. If a gray horse does not have a gray parent, then the purported parentage is likely to be incorrect (Trommershausen-Smith *et al.* 1976a).

A foal with two gray parents has at least a 25% chance to be homozygous for gray. Usually breed registries do not follow the breeding records closely enough to distinguish which grays are homozygous, but homozygous grays should only have gray offspring. When an apparent exception occurs in the studbook, usually it is found that the horse is gray, but the owner did not recognize the color when filling out registration forms and failed to notify the studbook in time to change the published color.

In at least one case, a possible example of transmissible back-mutation of gray to not-gray was identified. No evidence to deny assigned paternity was found for a single bay offspring of a homozygous gray Arabian stallion (with more than 200 foals) using 24 genetic systems with a combined efficacy of about 99% to detect incorrect paternity. Bloody shoulder marks may indicate back-mutation of gray in a clone of somatic (non-reproductive) cells.

The Thoroughbred studbook and racetrack nomenclature for grays is genetically confusing—grays born chestnut are registered as roan, while grays born dark bay, brown or black are registered as gray. Bays with G may be called roan or gray, depending on how conspicuous is the red hair component. (The classic roan gene, *per se*, is not found in Thoroughbreds.)

Breeds that have a high percentage of gray horses seem to have fewer flecked grays than breeds in which gray color is uncommon. No genetic analysis of the inheritance of the flecking trait in grays has been reported, but it is likely that a homozygous gray is of the clear white type. The flecked grays seem to be heterozygous. Grays without flecking need not be homozygous and in breeds in which gray is rare, usually they are heterozygous. A non-flecked heterozygous gray bred to a *gg* (not-gray) may have flecked gray off-spring. This situation suggests that flecking is a trait separate from gray, but as yet we have not identified how this trait is expressed in colors other than gray.

In dogs, a dominantly inherited trait for ticking puts dots of color in patches of white hair. Occasional horses have dots of color in the conventional pink-skinned white markings of the legs and face that may be produced by a horse gene homologous to the gene for ticking in dogs. The gene for flecking in grays is clearly different from a ticking trait since the color specks occur in the black-skinned areas, not in the white markings.

Gray interacts epistatically with all other coat color genes except white, obscuring their actions so that it is difficult or impossible to determine by looking at a gray horse what other coat color genes it has.

The gene linkage group for gray is currently undetermined.

Gene homology

A progressive graying gene present in some dog breeds, inherited as a dominant, is possibly homologous to the horse gene.

Breeds

Gray occurs in breeds throughout the world, including ponies, riding horses and draft horses. It is the predominant, but not exclusive, color in a few breeds (e.g. Andalusian, Kladruber and Lippizaner).

Defects associated with gray

Melanomas are more prevalent in gray than not-gray horses. The tumors are most commonly seen around the tail or the head, but may affect any organ system. Approximately 95% of melanomas in horses are benign, but may be disfiguring (Sundberg *et al.* 1977). If malignant they may cause organ dysfunction and lead to death.

Grays occasionally show skin depigmentation around the eyes, mouth and anus that may be considered unattractive, but is not a health risk. This can be a frustration to owners because in some horses the condition disappears and in others it persists. As with melanomas, depigmentation is also occasionally seen with other colors, but is more often associated with gray.

Roan

Roan produces a silvering effect by mixing white and colored hairs, generally more so on the body than the head and lower legs (Figure 21). The roan effect does not progressively whiten with age as does gray, although often the summer coat appears lighter than that of winter. Hair regrowth in areas of skin wounds may not show the white hair mixture, thus accentuating the appearance of scars (and brands) in the roan coat.

Figure 21: The interspersed mixture of light and dark hair on the body, but not the head or lower legs, characterizes the pattern produced by the roan (*RN*) gene.

A wonderful array of names can be used for the color variations produced by the combinations of roan with the basic colors, but most breed registries limit the options. In some schemes, "blue roan" may be used as the color term for black, brown or bay with roan and "red roan" for sorrel or chestnut with roan. Sometimes bay with roan is called "strawberry roan." Other breeds simply register the horse as "roan," losing the record of the basic coat color.

The presence of the roan allele is obvious by its silvering effect on coat color, and its inheritance follows a dominant pattern. However, a roan-type effect apparently can be produced by other genes, sometimes creating confusion in color designations for registration and in assigning genotypes. For example the gene responsible for leopard (appaloosa) spotting may be expressed as a mottled roaning effect, without distinct spots.

Trait inheritance and gene symbol

The roan trait is inherited as a dominant (*RN*). It has been described as a homozygous lethal from a studbook study (Hintz & Van Vleck 1979), although homozygous roan stallions have been reported (Geurts 1977).

Roan:	*RNrn* (*RNRN*?)
Not roan:	*rn rn*

"Roaning"

Many horses have at least a few scattered white hairs, seldom confused with the actions of *RN*, but occasional horses have a heavy dose of roaning—without having a roan parent to contribute an *RN* gene. In Arabian horses, among over 500,000 registrations in the AHRA studbook, 290 horses are designated as roan. Some of the roans are probably misidentified grays, particularly in early records. Considering only the roans without a gray parent, to eliminate records with the possible confusion of roan and gray color assignment, only 14% (12/85) of studbook recorded Arab roans have a roan parent. Using the traditional definition of roan as a dominant gene, these data could be taken to suggest a very high percentage of pedigree error among Arab roans or a significant under-reporting of the roan pattern. Parentage verification through genetic marker testing provides validation of the recent studbook records and there is no compelling evidence to support the notion that parentage assignment would be grossly inaccurate among the older records. If under-reporting occurs, it is not likely to be deliberate because most breeders appreciate the traditional status of this pattern and welcome its distinctiveness. Under-reporting is more likely due to difficulties in definition brought about by the assumption that extensive interspersed white hairs in the coat of an Arabian are caused by the classic roan gene and should be defined with the same criteria.

Compared with the classic roan gene, the roaning trait is typically an uneven pattern, heavier on the flanks and barrel than the forehand. Other prominent features include white flecking (irregularly shaped, white spots) on the flanks and belly, between the front and hind legs, and on the sides of the neck near where it joins the head. Particularly large flecked areas may be underlain with pink skin. The mixed coat may have a diffuse vertical white striping pattern reminiscent of brindling in dogs or cattle. Often the hair on the top of the tail is white, perhaps with several rows of prominent stripes across the top of the dock. The roaning trait is not confined to Arabians, but is found in many breeds including Thoroughbred and Quarter Horse. The inheritance of "roaning" has not been defined.

Gene homology

Roan is a recognized color trait in several species, but homologies among the various occurrences have not been studied. Roan color in cattle is caused by a gene that shows incomplete dominance (homozygotes are white). White females may have fertility problems. Roan color in mice is linked to piebald (*s*) on chromosome 14. Homozygotes are viable and fertile.

Linkage

Roan is part of LG II along with the white spotting gene tobiano (*TO*) and black pigment extension (*E*) (Andersson & Sandberg 1982).

Breeds

The *RN* allele is found in such diverse breeds as Quarter Horse, Peruvian Paso, Paso Fino, Welsh Pony, Miniature and Belgian, but is not found in many others, for example, Thoroughbred and Arabian. Roan is used for the assigned color in studbook and racetrack descriptions of some Thoroughbreds to distinguish browns or bays with the gray gene from chestnuts with gray—contributing to confusion about the definition of the roan gene. Roan is recognized as a color in the Arabian studbook, but in this breed an extensive interspersed white hair pattern is probably due to another gene or genes.

Chapter 5
Tobiano

The inheritance pattern of tobiano white spotting is well known to be a dominant trait. A tobiano foal must have a tobiano parent. Fillies and colts inherit tobiano from either sire or dam, or both. The tobiano gene is absent in the predominant North American breeds Quarter Horse, Thoroughbred, Standardbred and Arabian, but is found in a rich variety of breed types including Paint, Pinto, Dutch Warmblood, American Saddlebred, Tennessee Walking Horse, Missouri Fox Trotter, Paso Fino, Icelandic, Shetland and Miniature. Our discussion will focus on data from tobiano Paint horses, but the information about trait inheritance is equally valid for the tobiano pattern in other breeds.

Pattern description

The overall impression of a tobiano is that of a white horse on which large colored patches have been placed (Figure 22). The colored areas generally include the head, the chest and the flanks. Pink skin underlies the white areas and black skin is under the colored areas. The eyes are usually brown, but one or both may be blue or partially blue. The tail may be of two colors (white with black or red), a characteristic seldom seen in horses except for tobianos. The tobiano pattern is obvious at birth. Comparison of foal and adult photographs shows the large pattern definition remains constant during the horse's lifetime, although markings' outlines may change in small details. To be registered as a tobiano Paint, a horse must meet the American Paint Horse Association (APHA) pedigree and white markings requirements. The APHA describes the characteristic pattern features of tobiano as follows:

- The tobiano normally exhibits white across the spine, extending downward between the ears and tail in a clearly marked pattern.
- Head markings will be like those of a solid colored horse—solid or with a blaze, strip, star or snip.
- Generally, all four legs will be white, at least below the hocks and knees. The tobiano rarely has more than one or two solid-colored legs.

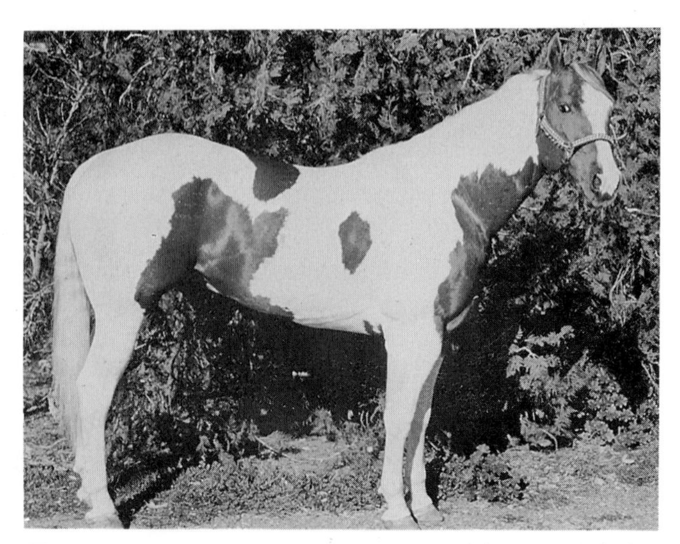

Figure 22: The pattern of white markings known as tobiano can be associated with any basic color. This Paint stallion shows the combination of tobiano and chestnut (*eeTOTO*).

- Generally, the spots are regular and distinct as ovals or round patterns that extend down over the neck and chest, giving the appearance of a shield.
- The horse will usually have the dark color on one or both flanks.
- A tobiano may be either predominantly dark or white.

Combination with colors and other patterns

Tobiano can occur with any coat color (sorrel tobiano, bay tobiano, palomino tobiano, dun tobiano, black tobiano and so on) and can occur in a mixture with other coat color patterns. Tobiano/overo and tovero are words used to describe the combination of tobiano with overo. Together these two patterns usually result in a horse with more white than colored area. Genes that produce leg and facial markings probably also interact with tobiano to affect the extent of white. A tobiano with minimal white pattern probably lacks the commonly found (and independently inherited) genes for white markings.

Tobiano can also occur with roan and with appaloosa spotting patterns. Since neither the Paint nor Appalossa breed allows registration of tobiano/appaloosa pattern blends, the "pintaloosa" is perhaps the best known today as a color variant in Miniature Horses.

Mule breeders know that the best way to get fancy stockings on a mule is to use a tobiano mare. For mules with the tobiano gene the extent of body spots may be quite restricted compared with the expression of tobiano in the horse parent, but the leg markings part of the pattern is consistently present.

Gene symbol

The genotype for a horse with the tobiano pattern is either *TOTO* (homozygous) or *TOto* (heterozygous). Deciding between the two genotype assignments requires putting together bits of information from pedigree, phenotype and get or produce records. Horses without the tobiano pattern gene are symbolized *toto*. No compelling reasons suggest that the genetic symbol for tobiano should be adopted from the nomenclature assigned to a spotting trait in another species, although if homologies are ever demonstrated, it would be prudent to use a uniform nomenclature.

Tobiano:	*TOTO* or *TOto*
Not tobiano:	*toto*

Worldwide distribution of tobiano

The tobiano pattern occurs in breeds worldwide, although "pied" is the English language term probably more often used outside the Americas. "Tobiano" appears to have been coined in South America, where it was traditionally used for distinctively spotted horses said to have descended from those brought by a Dutch emigrant named Tobias. In Britain tobiano horses may be called "painted," "colored," "piebald" (white and black), "skewbald" (white and any single color but black) or "odd-colored" (white and two or more colors), without distinguishing any particular pattern.

Besides the North American horse breeds with tobiano spotting listed at the beginning of this chapter, others worldwide include East Prussian Trakehners, and native ponies such as the Pottok from the Basque region of Spain and the Mongolian pony of central Asia.

Historical origin of tobiano

The distinctiveness of the tobiano pattern and its absence in many breeds worldwide tempts the thought that the tobiano gene could have a single historical origin. No documented account suggests that a tobiano horse has ever appeared in a breed which failed to include tobiano already. The origin of tobiano spotting is undefined in written history. Artwork from Asia suggests an ancient residence there for horses of tobiano pattern, but tracing the geographical origin of tobiano through existing art is difficult because many cultures lack significant historic collections of representational art. The historic sources of tobiano for Paint, Pinto, American Saddlebred, Tennessee Walking Horse, Missouri Fox Trotter, Shetland and Miniature breeds in North America may be unrecorded, but are assumed to have arrived as imported colonial horses from Europe.

With the development of DNA tests for horses, following their successful application to answer similar questions of gene origin for human diseases, in time it may be possible to know whether tobiano horses trace to a single progenitor. It is unlikely that origin could be taken to a specific breed without an ambitious DNA sampling of breeds of the world. The linkage phase association data already available argue for a single origin of most Paint tobianos. Whether horses that do not show the common linkage phase type are of independent mutational origin or arose by mutations from the common type, is yet to be determined.

Homozygous tobiano pattern?

Tobianos often have a few, small, colored spots in white areas but occasional horses show a dramatic proliferation of small, clustered spots often with roan edges, "halos," or roaning of the entire spot. Clustered colored spots that appear to be breaking through in otherwise large areas of body white have been called "ink spots" or "paw prints" (Figure 23). Horses that breed true for

Figure 23: Secondary spotting in white areas of tobianos are variously called ink spots or paw prints.

tobiano commonly have these spots, particularly horses with a moderate extent of white area. Although dramatically evident at times, the association of ink spots with homozygosity does not appear to be absolute. Their presence may be suppressed by genes that increase the extent of white area. Other (undefined) genes may produce small spots in white areas of any tobiano, even heterozygotes.

A true-breeding tobiano has two copies of the spotting gene (is homozygous for tobiano). At first it may be hard to comprehend that the pattern of tobianos with two gene copies may differ from that of single copy horses since breeders know that gene dosage for other dominantly inherited color variants (such as gray, bay and dun) cannot be determined by looking at the horse. A conspicuous and familiar example of differences between homozygous and heterozygous gene action occurs with the color dilution gene responsible for producing palominos. Chestnut is lightened to palomino or cremello depending on whether the dilution gene is present in single or double copy.

The tobiano pattern has traditionally been defined from single gene (heterozygous) horses. A modification of the tobiano definition to include the secondary spots seen in tobiano homozygotes may be in order. The presence of the extra spots provokes many unanswered questions: 1) what is the mechanism that produces them—are they evidence of back-mutation of tobiano to non-tobiano? or release of cells from inhibition of pigment production? or altered cell migration patterns? 2) why does it usually take two genes to produce these additional spots? Perhaps when we have information about the mechanism of tobiano spotting at the cellular level, we will be in a better position to answer these questions.

APHA studbook data

Tobiano traditionally is known to be inherited as a dominant trait, but it is worth examining the APHA studbook to verify this assumption. The occurrence of homozygotes and the typical trait transmission by heterozygotes should be confirmed. Other points to address with studbook data include the minimal expression of tobiano; evidence for back-mutation (gene instability) to the unspotted gene trait in records of homozygous tobiano stallions; the relationship of tobiano and overo spotting genes; and the involvement of tobiano in mostly white and in rarely-encountered viable white horses.

In the genetic analysis of studbook data, tobiano stallions are selected because mares cannot provide the numbers of offspring that the analysis requires for statistically significant conclusions. Test cross data, from breedings in which the mare was a solid colored QH or TB (homozygous recessive for non-tobiano, *toto*), provide the information for genetic analysis of the trait.

Homozygous tobianos

The records of six stallions likely to be homozygous for tobiano provided data for 835 foals from QH and TB mares. If these stallions were homozygous, all their foals in the test cross situation should be tobiano. Of the 835 foals, 830 were registered as tobiano (or tobiano/overo), four as overo and one as solid. Clearly this is subtantial evidence that the stallions were homozygous for tobiano. The small number of exceptions is not sufficient to refute their homozygosity, but can we explain them on another basis?

From pictures, three of the four overos were extensively white, probably over 90–95% of the body. The head was substantially white, but color occurred on the ears and poll, also on the chest and flanks. This pattern can be produced by various gene combinations, probably both with and without tobiano. The definition of genotype from phenotype for an extensively white (medicine hat) horse is difficult, but a better and quite reasonable pattern description of these three foals would probably have been tobiano/overo, not overo, particularly since they had a tobiano sire.

The fourth overo foal was clearly an overo, about 20% white. While no definite conclusion could be reached about this horse without further study, three possibilities can be considered, listed in order of most to least likely.

1. The reported parentage is incorrect. From studies in other horse breeds, pedigree error is about 0.5%, or 1 in 500 registrations. Perhaps closer examination of farm breeding records for the dam would provide evidence of cover by a stallion other than the homozygous tobiano. Alternative parentage options could be investigated with genetic marker studies if the horses are still alive and the owners and the APHA wished to pursue the question.
2. The tobiano gene is present, but not expressed. For this case, possibility 2 is less likely than 1, based on markings of known examples with tobiano minimal expression. Rarely, a horse with the tobiano gene will not have enough white to meet registry requirements. The white leg markings on minimally marked tobianos have a distinctive pattern of dark spots or streaks within the markings extending upward from the coronet. Also, hind leg markings above the hocks have a characteristic flat (horizontal) ending, compared to the tendency in non-tobianos of such markings to run up the anterior surface of the leg. Facial markings are usually not extensive. Breeding Stock horses with the tobiano gene might be designated "crypto-tobiano", to distinguish them from Breeding Stock that have not received a spotting gene. Several examples of minimal tobiano expression have now been studied and confirmed with blood marker analysis. In no case has a crypto-tobiano failed to have white leg markings.
3. There has been a mutation to overo. Since tobiano and overo are known to be the effects of different genes (see later section), this option is effectively proposing two mutational events. One event would be mutation of tobiano back to its non-tobiano alternative. The second event would involve the mutation of non-overo to overo. Individual mutational events are known to be rare. Since this explanation of an overo foal from a homozygous tobiano stallion requires two such events, the choice is definitely the least likely among the three proposed.

The fifth exceptional foal from the homozygous tobiano stallions was a mare registered as solid. Further investigation showed her to be a minimally marked tobiano. Her tobiano foal by a non-tobiano stallion (the foal's parentage qualified by blood typing) confirmed her crypto-tobiano status. Rare examples that at first appear to show tobiano skipping a generation, unexpected behavior for a dominant gene, instead provide examples that allow definition of the minimal expression of tobiano. A crypto-tobiano is genetically tobiano, but fails to meet the APHA phenotype definition for tobiano.

Some owners might favor a modification of the tobiano definition so that crypto-tobianos could be admitted to full registry status. A prudent strategy would be to fund research to develop a DNA-based diagnostic test for the tobiano gene. Owners of Breeding Stock horses might then have the option to petition for their full registry status based on the results of a direct gene test.

Back-mutation to non-tobiano is another possibility to explain solid colored offspring of a spotted parent, but no example of that effect has been demonstrated. Such an event is likely to be extremely rare, but if secondary spotting is evidence of tobiano gene instability in skin tissue, that instability could reasonably occur in the germ line tissue as well.

Heterozygous tobianos

A second aspect of confirming the dominant hypothesis for tobiano inheritance is to look at records of heterozygous stallions (e.g. stallions with tobiano dams and QH sires) bred to solid color mares. According to the genetic model (Figure 24), records should show 50% tobiano foals of both sexes from such matings, although if breeders do not register all the solid foals produced, the percentage of tobianos will be higher.

Genetic contribution from mares	Genetic contribution from stallions	
	TO	*to*
to	*TOto* tobiano	*toto* solid
Offspring proportion	50%	50%

Figure 24: Foals resulting from test cross breedings of heterozygous tobiano stallions are predicted to be either tobiano or solid, in equal numbers.

Five heterozygous stallions together had 303 foals out of solid colored (QH or TB) mares, of which 201 (66%) were registered as tobiano. This is higher than the 50% anticipated but not significantly different by statistical testing from the dominant gene model.

Linkage group for tobiano

The gene for the tobiano trait belongs to one of the genetic linkage groups currently identified for the horse. A linkage group is a series of genes that tends to be inherited together. Such genes are physically close together on the linear genetic information structures known as chromosomes. The horse linkage groups are assigned numbers in order of their discovery. Tobiano is in the second linkage group recognized for the horse, designated LG II (Trommershausen-Smith 1978). Since LG II has not yet been assigned to a particular chromosome, it may also be called U2 (unassigned group 2).

In close linkage with tobiano are two genes that code for blood proteins and two genes that affect coat color (Andersson & Sandberg 1982, Bowling 1987, Sandberg & Juneja 1978, Sponenberg *et al.* 1984). The genes albumin (*ALB*) and group-specific component (*GC*) are so closely linked to tobiano that no recombination has been detected in the families studied. The relative gene order of *ALB*, *GC* and *TO* is therefore not known. The other two genes affecting coat color in LG II are black pigment extension (*E*) and roan (*RN*). *E* is the basic gene producing coat color differences in horses. The homozygous recessive (*ee*) horse is unable to produce black pigment in hair so will be a chestnut, sorrel, red dun or palomino. Bay, black, grulla and buckskin horses are *EE* or *Ee*. *RN* is a dominantly inherited pattern gene that causes white hair to be interspersed throughout the body coat, generally leaving the lower legs and head without sprinked white. *RN* and *TO* are linked through their common association with other genes in LG II, but their direct, genetic interaction has not been studied.

Besides the five genes already mentioned, the linkage group contains three other blood protein genes: haptoglobin (*HP*), esterase (*ES*) and the mitochondrial form of glutamate oxaloacetate transaminase (GOT_m) (Andersson *et al.* 1983b, Weitkamp *et al.* 1985). Only limited data exist for *HP* and GOT_m. Conventional blood typing tests provide information about *ES* markers, but the gene is too far from the *ALB–GC–TO* complex to be practically useful for tobiano breeders.

Phase association of LG II markers in tobianos

Nearly all tobiano marked horses—from a variety of breeds—have the same set of markers for the closely linked blood proteins, namely, ALB-B and GC-S (Bowling 1987). These markers are not unique to tobiano horses and can be found in all horse breeds, including those without tobiano spotting. Breeds so far studied that demonstrate the tobiano linkage phase conservation include American Saddlebred, Icelandic, Miniature, Paint, Pinto, Paso Fino, Shetland, Missouri Fox Trotter and Tennessee Walking Horse. Not only is the phase conservation of tobiano with a set of protein markers unexpected, but the range of breeds is not anticipated. The conserved phase association among tobianos of these breeds may provide a clue about the breeds' historical connections, or at least about the tobiano horses within these breeds.

Genetic theory predicts that the tobiano marker would be distributed among the eight possible classes of *ALB*, *GC* and *E* markers in proportion to the combined relative frequencies of the alleles at these loci. Population data (see Chapter 16: Genetic descriptions of breeds) show that *ALB* markers are not randomly distributed in tobiano horses. The high frequency of GC-S among tobianos is a conspicuous departure from expectation, since it is an infrequent marker in most breeds.

In Paints at least 90% of tobianos have the ALB-B and GC-S factors (Table 2). Gene linkage could provide skewed color class frequencies within

Markers			Frequency of marker phase association in Paint horses
ALB	*GC*	*E*	
B	S	E	Common
B	S	e	Common
A	F	E	Infrequent
A	F	e	Rare
B	F	E	Infrequent
B	F	e	Rare
A	S	E	Very rare
A	S	e	Very rare

Table 2: Phase association frequency for *ALB*, *GC* and *E* gene markers of tobiano Paint horses.

families, but would not be expected to be significantly evident in a large population or breed unless some mechanism holds the linkage group together so tightly that the normal processes of genetic recombination during meiosis are suppressed.

The gene complex involving tobiano may be held together in a special chromosome arrangement called an **inversion**. If the inversion can hold the linked genes together so strongly that changes are virtually non-existent, then **founder effect** may explain the unequal distribution of classes in Paints. That is, the very rare individuals in which the association was modified historically have made only a small genetic contribution to the Paint breed, compared with the more common type. This situation could change if a stallion with the infrequent phase becomes extremely popular, but it would still take a long time to upset the overall numerical dominance of the *ALB-B, GC-S* type in tobiano Paints.

Additional indirect evidence to support the inversion hypothesis is provided by color/pattern data (discussed next), but direct proof awaits further research. Whatever the underlying mechanism, the tobiano pattern is clearly associated with an unusual genetic situation.

Recombination of LG II genes in tobianos

The parents' *E* and *ALB–GC* linked marker combinations are passed to nearly every offspring whether or not the matings involve tobiano. Recombination between markers of the parents' linked genes will normally occur in a few offspring. The expected rate of recombination is in proportion to the linear chromosomal distance between genes. For non-tobiano horses the genetic distance between *E* and *ALB–GC* is 7 cM, which predicts 7% of offspring of tobianos will also be recombinant for these genes. In tobianos the recombination process between *E* and *ALB–GC* appears to occur much less frequently than predicted from non-spotted horse data. Recombination occurs occasionally, but it appears to be rare. The reduced combination frequency between *E* and *ALB–GC* in tobianos is indirect evidence supporting the suggestion that the tobiano gene is involved in a chromosomal inversion.

In certain situations breeders can clearly see the effect of close linkage between tobiano and color genes. Remember, gene linkage causes the marker combinations transmitted by parents to offspring to stay together in the associated phases of the parents. For example, when bred to sorrel mares, the spotted foals of a bay tobiano stallion with a sorrel QH (solid) dam will be bay like the sire and the solid foals will be sorrel (Figure 25).

Sorrel tobiano and solid bay offspring are the rare classes found in this particular breeding combination. If tobiano and color were not linked, equal numbers of offspring in four color/pattern classes would be expected. Since *RN* is also in LG II with *TO*, some stallions with both *RN* and *TO* should transmit the genes together and others should transmit them separately. Roan tobiano Paint horses are not very common. It seems that *RN* and *TO* genes are on different homologs of the chromosome pair in these cases and would therefore be transmitted as though they were alternatives. A recombination event could cause the two genes to be on the same chromosome and in other breeds (such as Miniature) this combination may be found.

Use of phase association in marker-assisted selection

How does information about gene linkage of tobiano and its unusual behavior benefit Paint horse breeders? The best way to follow a gene in pedigrees would be with DNA analysis, but the DNA sequence of the tobiano gene is not known. Markers for the linked genes associated with tobiano may be used to track the tobiano chromosome through pedigree generations until a direct gene test becomes available. Routine blood typing tests required by the

Genetic contribution from mares	Genetic contribution from stallion			
	E TO	*e to*	*e TO*	*E to*
e to	*Ee TOto* bay tobiano	*ee toto* solid sorrel	*ee Toto* sorrel tobiano	*Ee toto* solid bay
Offspring proportion	about 50%	about 50%	RARE	RARE

Figure 25: Color/pattern classes of offspring from bay tobiano stallion (*Ee TOto*) bred to solid sorrel mares (*ee toto*). The table shows that the effect of gene linkage is to change the expected ratios, effectively reducing the number of offspring classes from four to two.

APHA for breeding stallions currently provide genetic markers for ALB and GC systems. Analysis of this trio of genes allows breeders the option to apply marker-assisted selection to identify homozygous tobianos. Blood testing and marker analysis could provide genetic information about young foals before proof of zygosity could be obtained from breeding results. An example of this analysis process is given in a later section.

Genetic relationship between tobiano and overo

A test cross can clarify the genetic relationship of tobiano and overo. If tobiano and overo were alleles, tobiano/overo (tovero) stallions bred to QH and TB mares would produce only tobiano or overo foals, no solids and no toveros. Studbook data from such horses counter that proposal, invariably showing four pattern classes of foals, in approximately equal numbers. For example, one tovero stallion in 42 breedings to QH mares sired nine solids, ten overos, ten toveros and 13 tobianos. These data show that separate, unlinked genes produce the tobiano and overo patterns (Figure 26).

Tobiano/overo horses may be nearly white (medicine hat) showing an additive effect of the two traits' gene actions. Interaction of tobiano and overo genes apparently can produce a white horse, although among Paint horses it is probably more common for a white foal to be the product of two overo parents and to die within hours or a few days of birth. However, rare viable white Paints occur and breeding records show them to have foals of the four pattern classes (Figure 26), though not necessarily in equal numbers.

Predicting the likelihood of a tobiano foal by a tobiano stallion

To evaluate the pattern breeding potential of a tobiano stallion it is his offspring from solid mares that tell the tale. Typically a tobiano stallion is heterozygous for tobiano so 50% of his foals from solid mares will inherit his

Genetic contribution from mares	Genetic contribution from stallion			
	O TO	*o to*	*o TO*	*O to*
o to	*Oo TOto* tovero	*oo toto* solid	*oo TOto* tobiano	*Oo toto* overo
Proportion	25%	25%	25%	25%

Figure 26: Foals of a tovero stallion in test cross matings show independence of genes for tobiano and overo, with equal numbers of offspring in four classes.

tobiano pattern gene. If a tobiano mare is bred to such a stallion, then the foal has a 75% chance of being spotted. If the mare is overo, then the foal also has a 75% chance of being spotted—either tobiano, overo or a combination of tobiano and overo. A few tobiano stallions and mares are homozygous for the tobiano spotting gene and all their foals are expected to inherit the tobiano pattern, regardless of the spotting characteristics of their mates.

Tests for homozygosity

The minimum requirement for a true breeding tobiano is a tobiano sire and a tobiano dam. A true breeding tobiano can be obtained—with an abundance of luck—in one generation. One-fourth of the offspring from matings of heterozygotes will be solid, not tobiano, but among the tobiano offspring, one-third could be homozygous for tobiano (Figure 27).

Some experienced tobiano breeders believe tobiano × tobiano breedings do not produce homozygous tobianos as frequently as predicted. Data are not available to verify the prediction, but hopefully this point can be substantiated soon, given the current interest in homozygous tobianos among Paint breeders.

If a homozygous tobiano of a specific sex and color is desired, it would be wise to arrange several matings to assure the likelihood of success. A homozygous tobiano *colt* would be expected in 12.5% of tobiano × tobiano matings (one in eight). A *buckskin* homozygous tobiano colt would occur in only selected matings and in those the frequency is highly unlikely to be greater than 6.25% (one in sixteen).

Evidence for tobiano homozygosity is based on assembling as many as possible of the following bits of evidence, which will be discussed in order:

- **Pedigree:** both parents tobiano (presumably the parentage is correctly recorded).
- **Phenotype:** tobiano with secondary spotting.

Genetic contribution from mares	Genetic contribution from stallions	
	TO	*to*
TO	*TOTO* tobiano 25%	*TOto* tobiano 25%
to	*Toto* tobiano 25%	*toto* solid 25%

Figure 27: Predicted classes and proportion of offspring in matings between heterozygous tobianos. Homozygous tobiano class is shown in darker shading. One-third of the tobianos are expected to be homozygous.

- **Test cross breeding:** no solid color offspring from a sufficient number of solid color mates.
- **Studbook data:** no solid color offspring.
- **Genetic marker analysis:** diagnosis of linked marker genes (*ALB* and *GC*) and tracing inheritance of tobiano chromosome in pedigree through analysis of records of parents and offspring.
- **DNA test:** demonstration of two copies of tobiano DNA sequence (not currently possible since the gene has not been characterized at the DNA level).

Pedigree

To be true breeding (homozygous) for tobiano, a horse must have received a tobiano-bearing chromosome from each parent. Thus both parents must be tobiano (but not themselves necessarily true breeding).

Phenotype

Whether the homozygous tobiano always has a distinctive spotting pattern distinguishable from the heterozygotes remains to be determined, but numerous examples suggest the occurrence of multiple secondary spots in white body areas is strongly correlated with homozygosity for tobiano.

Test cross breeding

Homozygosity can be demonstrated with foals from test cross breedings of a poential homozygote to solid mates. A solid foal from any mating is presumptive evidence to exclude homozygosity for tobiano (but rare exceptions could be accommodated in an extensive record); two or more solids would be difficult to attribute to any hypothesis but heterozygosity. The statistical assurance of homozygosity is related to the number of offspring available (Table 3).

Studbook data

When breeders choose not to register solid foals (as Breeding Stock), the studbook record may be an inadequate validation for a zygosity determina-

Number of tobiano offspring, no solids, from matings of tobiano to solid	Prediction certainty for tobiano parent to be homozygous
5 tobianos	97%
7 tobianos	99%
10 tobianos	99.9%

Table 3: Statistical assurance of tobiano homozygosity prediction from test cross breeding (tobiano × solid). Accuracy of prediction is related to number of offspring.

tion, particularly when the number of offspring is relatively small. However, a stallion with a studbook record of 50 tobiano foals and no solids is certainly an excellent candidate for homozygosity.

Real-life complications include the occurrence of a single solid foal in such a record. Several hypotheses can allow this record for a homozygous tobiano, including that the solid foal 1) has incorrectly recorded parentage, 2) is a crypto-tobiano, or 3) is an extremely rare back-mutation to non-tobiano (yet to be convincingly demonstrated). Often it is impossible to undertake the research necessary to reconcile the genetic basis underlying the apparent problem record. A good understanding of genetics allows one to weigh the studbook data in light of the other bits of presumptive evidence one can assemble to ascertain homozygosity for tobiano.

Genetic marker analysis

For mares and young stallions neither test breeding schemes nor studbook records may provide adequate information for tobiano zygosity analysis. Genetic markers linked to tobiano may allow a tobiano zygosity prediction if the particular variants of the blood proteins that are traveling on the same chromosome as the tobiano gene can be determined for the sire and for the dam. If a foal receives each parent's tobiano chromosome, the offspring will sire or produce only tobiano foals. The problems with tests based on analysis of linked markers must be clearly understood by anyone planning to use them as a tool to select breeding stock:

- Some matings are uninformative, but having two linked markers (for tobiano this means *ALB* and *GC*) makes this possibility less likely than if only one marker was available.
- Misdiagnosis is possible due to inability of the tests to detect rare recombinants.
- In some families, if parents or other offspring are unavailable to assign or confirm linkage assignments, the test has low informativeness.

An example of the analysis of blood typing information and coat color to predict the gene dosage (zygosity) of a tobiano foal who is the offspring of tobiano parents is provided (Table 4).

Foals A1, A2, B1, B2 and C1 are all offspring of stallion S. Foal C2 is by an untested stallion. Foals B1 and C1 are the offspring whose tobiano zygosity is in question. To predict whether B1 is homozygous or heterozygous for tobiano requires blood typing information from its dam (mare B) as well as other offspring of the stallion, both solid and tobiano. These predictions will be based on data interpretations that assume no recombination between any markers in the *ALB–GC–TO* complex. Mare A is extremely useful for this analysis. She is homozygous for all four genes and her foals allow us to define the gene linkage phases (haplotypes) for each of the stallion's chromosomes. The solid foal A1 of phenotype ALB-A, GC-F is presumably of genotype *ALB-AA, GC-FF*. It must have inherited ALB-A and GC-F from the stallion. The tobiano

	ALB	*GC*	Coat color	Pattern
Stallion S	AB	FS	Bay	Tobiano
Foal A1	A	F	Chestnut	Solid
Foal A2	AB	FS	Bay	Tobiano
Mare A	A	F	Chestnut	Solid
Foal B1	B	S	Bay	Tobiano
Foal B2	A	F	Bay	Solid
Mare B	AB	FS	Bay	Tobiano
Foal C1	B	S	Bay	Tobiano
(Foal C2)	B	S	Bay	Solid
Mare C	B	S	Bay	Tobiano

Table 4: Blood typing data (*ALB* and *GC*) predict foal B1 is homozygous for tobiano, but are inconclusive for predicting tobiano zygosity of C1. A single letter in the *ALB* or *GC* column is the phenotype of a homozygote. All foals except C2 are sired by stallion S.

foal A2 inherited ALB-B, GC-S and tobiano from the stallion, since these three markers could not have been transmitted by mare A. These two foals of mare A show that the stallion's solid chromosome has the haplotype (*ALB-A*, *GC-F*, *to*) and his tobiano chromosome has (*ALB-B*, *GC-S*, *TO*) (parentheses indicate that gene order is unknown). Foal B2 shows that mare B's solid (non-tobiano) chromosome has ALB-A and GC-F. Foal B1 inherited ALB-B and GC-S from both its sire and its dam and is predicted by pedigree and blood protein markers to be homozygous for tobiano.

The solid foal C2 shows that mare C is heterozygous for tobiano, but her *ALB* and *GC* markers are not informative for tobiano linkage testing. Her tobiano and solid chromosomes are indistinguishable by *ALB* and *GC* tests since she is homozygous for markers of both genes. Thus the blood protein marker tests for tobiano zygosity status are inconclusive for C1 and, indeed, for any tobiano foal of C by any tobiano stallion.

Due to the tight linkage of *E* to *ALB–GC–TO*, even from the limited data presented in Table 4, the stallion's chromosomes can be assigned the phases (*ALB-B*, *GC-S*, *TO*), *E* and (*ALB-A*, *GC-F*, *to*), *e*. From chestnut mares, nearly all the tobiano foals of this stallion can be predicted to be bay (or black) and nearly all the solid foals will be chestnut. The exceptions are predicted to occur due to infrequent chromosomal recombinant events between extension and the tobiano–albumin–GC complex.

Despite the drawbacks of linkage testing, it is proving useful to many tobiano breeders when extensive production data are unavailable, such as for mares and young stallions.

DNA test

The direct test for demonstrating which offspring are homozygotes would be a DNA sequence analysis for the tobiano gene regions of each chromosome.

Since such a test is not available, zygosity determination involves assembling other bits of information.

Genes similar to tobiano in other animals

Linkage relationships among homologous genes are highly conserved between different species. The research database for the mouse is a particularly good information resource for coat color mutants and linkage relationships. What dominant spotting genes in the mouse are linked to the blood protein genes *ALB* and *GC*? The mouse albumin gene is known to be on chromosome 5 and very closely linked are three spotting genes Patch (*Ph*), Dominant spotting (*W*) and Rump-white (*Rw*). These mouse genes differ from tobiano in that they are all lethal in the homozygous condition and may also be associated with defects including anemia and infertility that are not observed with the tobiano pattern of horses. In the rat and human, linkage is also observed between spotting genes and the *ALB–GC* complex. Clearly, several species may help to identify the DNA sequence of the horse tobiano gene. Likewise, gene mapping research in horses is expected to contribute to an understanding of genes in other organisms by providing examples of homologous gene variation and expression.

Defects associated with tobiano

No defects which compromise health have been identified to be consistently associated either with homozygous or heterozygous tobianos.

Breeding Stock tobianos

Solid color offspring of matings between tobianos or a tobiano and a solid parent can be registered as Breeding Stock with the APHA. A rare few may be minimally marked tobianos (crypto-tobianos), but most will be solid *because* they have failed to receive a tobiano gene. To produce a tobiano foal using a Breeding Stock horse will require a mate with the tobiano gene, since the Breeding Stock horse lacks it. Breeding Stock horses may be a source of valuable conformation- or performance-related genes and need not be eliminated from a breeding program, but they should not be expected to contribute tobiano pattern genes to an offspring.

CHAPTER 6
Overero

Occasional white splashed foals from modestly marked Quarter Horse parents are part of the Western livestock heritage. Horses with markings outside the bounds accepted by the American Quarter Horse Association (AQHA) cannot be registered as Quarter Horses, but may be registered as overos by the American Paint Horse Association (APHA) and the Pinto Horse Association of America (PtHAA).

The unexpected and unpredictable occurrence of spotted foals from registered Quarter Horses has been used as evidence for the inheritance of overo as a recessive trait. Paint horse breeders plan matings for overo assuming it to be caused by a recessive gene. However, transmission of overo spotting in Paint horses generally *does not* follow a recessive pattern. Data compiled from the APHA studbook show that many examples of overo follow the genetic pattern expected for an autosomal dominant trait. Paint horse breeders will want to be familiar with these findings and their implications for breeding program strategies.

Pattern description

The overall impression of an overo is that of a colored horse with white patches (Figure 28). Pink skin underlies the areas of white. The eyes are usually brown, but one or both may be blue or partially blue. The overo pattern is obvious at birth. Comparison of foal and adult photographs shows the large pattern definition remains constant during the horse's lifetime, although some markings' outlines may change, particularly on the belly. To be registered as an overo Paint, a horse must meet APHA pedigree and white markings requirements. The APHA describes the characteristic overo pattern features as follows:

- The white originates on the underside of the horse and will rarely cross the back of the horse between its withers and its tail.
- Generally, at least one, and often all four legs will be the dark color.
- Head markings are predominantly white; often bald, apron or bonnet-faced.

Figure 28: Irregular white markings on the neck and sides of the barrel of this bay Paint stallion illustrate overo. This particular pattern is sometimes called "frame-overo."

- Generally, the white is irregular, rather scattered or splashy. It is often referred to as calico.
- The tail is usually one color.
- An overo horse may be either predominantly dark or white. (The darker color is more common.)

Combination with colors and other patterns

Overo can occur with any coat color (sorrel overo, bay overo, palomino overo, dun overo, black overo and so on) and with other patterns. Tobiano/overo and tovero are words used to describe the tobiano and overo composite. A horse with genes for both patterns usually has more white than colored area. Genes that produce leg and facial markings probably interact with overo to affect the extent of white. Overo can also occur in combination with various appaloosa spotting patterns (Figure 29), but neither the Paint nor Appaloosa breed allows registration of overo/appaloosa spotting blends.

Overo pattern variants

Mammalian pigment cells (melanocytes) originate in the neural crest region of the early embryo and migrate during later developmental stages to populate selected areas throughout the body. The overo pattern could be a consequence of melanocytes failing to migrate or produce pigment, or of destruction of pigment cells after they reach certain areas. No studies have defined the status of melanocytes in overos, but it is assumed that white spots are probably a consequence of melanocytes failing to migrate to these areas.

Figure 29: A bay mare demonstrating the combination of overo and appaloosa patterns.

The term overo in the APHA studbook probably applies to genes at more than one locus. In mice some 25 genes control spotting patterns, so it would not be surprising to find that horses also have several genes for overo spotting. The differences produced by the various overo genes probably overlap in many phenotypic details. The classic way to characterize spotting genes in mice is to define their linkage relationships to each other and to standard panels of mapped genes. (This strategy has been useful for tobiano spotting in horses.) Linkage comparisons potentially provide a method to define the different overo patterns genetically, eventually leading to distinctive phenotype descriptions for the action of each gene. Techniques of molecular biology that define DNA gene sequences will ultimately provide the most accurate way to distinguish the variants, but these are not presently available.

Among Paint horses the most prominent and distinctive overo pattern is that sometimes called "frame-overo", in which the dark color typically occurs along the topline, chest, legs and tail with white occurring in a horizontal motif on the body, accompanied by substantial white face markings. Most of the discussion in this chapter is specific to this pattern. Other overo patterns in Paint horses that may be genetically distinct from frame-overo are "sabino" ("calico") (generally four white legs, jagged markings and extensive roaning) (Figure 30) and "splashed-white" (four white legs extending up onto the belly, bonnet or apron face markings, usually without extensive roaning) (Klemola 1933, Sponenberg & Beaver 1983). Another possible overo pattern type could be the maximal expression of the several "minor" spotting genes responsible for "conventional" leg and face markings.

These various overo patterns may be alleles of a single gene or may belong to several genes. Eventually the phenotype of Paint horse overo variants should be defined in three areas: the phenotypes of single and double dose combinations (heterozygotes and homozygotes) for each gene; the com-

Figure 30: Overo Paint mare showing the sabino ("calico") pattern.

bination of any two or more of the overo genes; and the combination of any of the overo genes with tobiano. For the most part these descriptions are not currently possible, but are obvious research goals.

Worldwide distribution of overo-type patterns

"Pied" may be the English language term more often applied outside the Americas to overo-type horses, but it may be used for tobianos as well. Overo ("bird's egg pattern") is the term used in South America, probably mostly for what is known in North America as the sabino type. In Britain "painted" or "coloured" horses are more generally split into the categories "piebald" (black and white), "skewbald" (white and any single color but black) or "odd-coloured" (white with two or more colors), without distinguishing between patterns.

Sabino, splashed-white and frame-overo type patterns seem to occur in breeds worldwide, although the sabino type is much more common than the others. Without attempting a comprehensive listing, breeds with at least occasional examples of extensive leg and facial markings patterns with roaning and splashes of body white (sabino pattern) include Thoroughbred, Arabian, Tennessee Walking Horse, American Saddlebred, Shire, Clydesdale, Welsh, Miniature and South American Criollo breeds. Sabino is sometimes considered to be an exaggerated version of the standard markings traits possessed by domestic horses, but sabino genes may be genetically distinct from genes for conventional markings.

The term overo is becoming widely but not necessarily appropriately adopted for any white spotting pattern that is not tobiano or appaloosa. Some owners of Arabians refer to body spots and tall stockings as overo pattern, overlooking that the term particolored has more traditionally been used for Arabians with such markings. Overo is not a markings description term applied by any Arab registry authority.

In the discussion to follow on the genetics of overo, the Paint Horse breed is the source of the specific data presented. The conclusions would also apply to Pintos. White spotting in other breeds may be produced by genes similar to those found in Paints and Pintos, but different genes may be involved as well.

Gene symbol

Overo spotting of Paint horses well illustrates the value of the letter system developed by geneticists nearly 90 years ago to symbolize hypothetical genes. Letters are assigned to genes based on results of trait transmission data, without knowing the chromosomal location or DNA sequence of the gene. The gene symbol *O* (capital letter symbol in italics) is assigned to the overo gene. The presence of spotting is inherited as a dominant trait.

Overo:	*Oo*
Not overo:	*oo*

The genotype for a horse with the overo pattern is *Oo* (a heterozygote). The *OO* genotype (a homozygote) is likely to be the cause of a known lethal condition. Horses without the overo pattern (*oo*) are homozygous for the recessive allele symbolized *o* (lower case italic). The same symbols have previously been hypothesized to act in the reverse direction, with the overo pattern produced as a result of the homozygous recessive condition. While it is possible that a few horses registered as overo have patterns inherited as recessive traits, the recessive model does not adequately account for the majority of the available Paint breeding data.

If gene homologies to spotting patterns in other mammals can be demonstrated, or if the developmental or physiological actions of the horse genes can be defined, the genetic designation for overo may be changed to reflect that information. For the general discussion of overo in this chapter, the gene symbol *O* is used for simplicity, but realizing that when better gene definition can be achieved, the symbol may be changed. Additional symbols for spotting may be added when the patterns can be reliably distinguished. The general assumption given here that overo is a homozygous lethal may not apply to all patterns currently called overo.

Genetic studies of overo

Classic Mendelian genetic research constructs matings in which one parent has the trait and one does not. If no offspring shows the trait, the alternatives to be considered include 1) the trait is not inherited (may be produced by an environmental effect), 2) the trait is recessive, requiring gene contributions from both parents, or 3) the trait has reduced penetrance (not always expressed even when the gene is present). While we cannot rule out any of these as possibilities for some types of overo, our research strongly suggests none fits the most common breeding scenarios of Paint horses.

Abundant evidence from the APHA studbook shows that a mating between one overo parent and one solid parent often produces an overo. (If it were unlikely that a spotted foal would be produced, horse owners would probably become disenchanted with a pattern-based registry that allows crossbreeding.) The data support the model of overo spotting in Paint Horses as a dominant trait, requiring only one parent to transmit the gene (Bowling 1994a).

In the tradition of classical Mendelian genetic analysis, studbook records for overo stallions bred to solid color mares (QH or TB) provide the test cross genetic data to evaluate the dominant gene model.

Cropout (overo) breeding stallions

The studbook record of a prominent splashed-white brown Paint stallion, the product of QH parents (a cropout), provides a starting point for the discussion. An APHA listing provided data on 178 foals, but it is the 39 offspring from non-spotted mares that give information most useful for genetic analysis. In matings to QH and TB mares he sired 20 overos and 19 solids.

If overo is due to a recessive gene, the stallion would have an overo gene pair and would transmit an overo gene to all his foals. His overo offspring must have received an overo gene from their solid colored dams as well. To have, on average, an overo gene transmitted to half his offspring from their QH (or TB) dams, logically implies *all* Quarter Horses (or Thoroughbreds) carry one overo gene. Following that reasoning, 25% of QH × QH matings would result in cropout foals, which is clearly not true for overo in general, or splashed-white in particular, probably the least frequent of the overo types.

A dominant gene hypothesis for overo provides a much better fit to the studbook data than does a recessive. The cropout overo stallion appears to contribute the pattern gene to half his foals (Figure 31). The dominant transmission pattern seen in the record of the splashed-white cropout stallion is found for other cropout examples, both of frame-overo and sabino types.

An overo cropout in a QH × QH mating might signal a single mutational event, affecting only one chromosome transmitted by one of the parents. The mutation is then passed on as a dominant gene by the cropout offspring. (The mutation part of the proposal will be discussed in more detail later.)

Stallions that are the product of overo × solid matings

Under the assumption that overo is inherited as a dominant gene, overo stallions with one overo parent and one QH or TB parent would be heterozygous for overo. Again, the studbook record overwhelmingly supports the hypothesis of overo as a dominant gene. When bred to solid color QH and TB mares,

Genetic contribution from solid mares	Genetic contribution from cropout stallions	
	O	*o*
o	*Oo* overo	*oo* solid
Offspring proportion	50%	50%

Figure 31: Dominant gene model for cropout overo stallions agrees with observations of 50% overo foals and 50% solid-colored foals from matings between solid mares and cropout stallions.

these stallions on average sire 50% overo foals and 50% solid colored foals, with occasional stallions siring more than 50% overos, but apparently never 100%.

Stallions that are the product of overo × overo matings

Stallions siring more than 50% overo foals in the studbook records could occur because foal owners elect not to register solid foals, as we found for heterozygous tobiano stallions. But more than 50% overo foals by stallions bred to solid mares could also be the result of overo stallions with more than one overo gene. If more than one type of spotting is registered as overo in Paints and the combinations are not lethal, stallions who are the offspring of two overo parents could have more than one overo-type gene. A stallion with one dominant gene would sire 50% spotted foals from solid mates (Figure 31); one with two different dominant genes would sire 75% spotted foals (comparable to the tovero breeding model (Figure 34) but with overo pattern substituting for the tobiano and tovero classes). A stallion with three dominant genes would be expected to sire 87.5% spotted foals from solid mares (Figure 32). This model should be regarded as highly speculative, because the number of genes involved in producing patterns called overo is not known, but some data are available to provide tentative support.

If predominantly white overo stallions arose from the interaction of genes at different loci, then these stallions would be good candidates to have studbook records showing more than 50% overo foals from solid mares. Data analysis for four substantially white stallions, the products of overo × overo breeding, provided useful examples to verify a model with multiple overo-type genes. One stallion had 26 overos among 36 foals from solid mares and another had 11 overos in 15 foals. Together they had 37 overos among 51 foals (73% overos), closely approximating the two-gene model (75%). Two other stallions together had 82 overos among 92 foals reported from solid mares (89% overos), approximating 87.5% expected by a model involving three dominant genes.

Genetic contribution from solid mares	Genetic contribution from stallion							
	KLM	*KLm*	*Klm*	*kLM*	*kLm*	*klm*	*klM*	*KlM*
klm	*KkLlMm* overo	*KkLlmm* overo	*Kkllmm* over	*kkLlMm* overo	*kkLlmm* overo	*kkllmm* solid	*kkllMm* overo	*KkllMm* overo
Offspring proportion	12.5%	12.5%	12.5%	12.5%	12.5%	12.5%	12.5%	12.5%

Figure 32: Model for test cross mating of overo stallions heterozygous for three independent (unlinked) dominant genes for white spotting. For simplicity the three different genes are symbolized *K*, *L* and *M*. Offspring having at least one dominant gene will be overo (87.5% of foals). The amount of white in the coat is expected to increase according to the number of dominant genes.

Matings between two-gene overo stallions and overo mares may produce a high percentage of overo foals, but lethal white foals (see next section) would be expected to occur more frequently than in breedings in which one parent was solid colored.

Homozygous overo

The record of overo × overo breedings provides additional evidence against the recessive overo hypothesis. If overo is recessive, overo × overo matings should breed true for overo, with no solid color foals produced. From the studbook record overo × overo matings produce both overo and solid colored foals in substantial numbers. One way to explain the solids is to propose overo may have reduced penetrance, that is to say, the gene is present but either is not expressed or its minimal expression falls outside the APHA definition. While reduced penetrance undoubtedly occurs with overo genes, that explanation is not sufficient to explain all the data associated with overo breeding.

A second way to explain the solid colored foals is to fall back on the hypothesis that overo is a recessive gene, adding the possible involvement of more than one locus. The mating of two overos each homozygous for different recessive genes would mostly produce solid (colored) offspring, heterozygous at each locus. Under this model, in an overo × overo mating the ratio of overo to solid offspring would depend on the carrier frequency for the "other" overo gene. Since the studbook record clearly shows many overo × overo matings produce overo offspring, the recessive-multiple-overo-gene hypothesis would require that overo have a high frequency in the solid population, not consistent with the overall rare occurrence of cropouts in QH breeding.

The recessive-multiple-overo-gene hypothesis is not sufficient to explain another observation about overos, outside the studbook record. Reports in veterinary medical journals describe foals, substantially white in color, that lack nerve cells in the intestine and live less than a week. These lethal white foals are primarily from overo × overo matings of Paint horses, although it is not always clear from the reports which of the possible overo patterns may be involved. No successful surgical correction of the neurological defect has been reported. Studbooks only record offspring that live long enough to be registered, so information on the mating types that produce defective offspring will necessarily come from research studies and owners' reports.

Taken together, published information from the studbook record and medical journals record three pattern types for foals of overo × overo matings: overo, solid colored and lethal white. The three classes fit a model of overo as a dominant gene with lethal effects in the homozygous condition. This model predicts that 25% of foals from two overo parents would be lethal white (Figure 33).

Genetic contribution from mares	Genetic contribution from stallions	
	O	*o*
O	*OO* lethal white 25%	*Oo* overo 25%
o	*Oo* overo 25%	*oo* solid color 25%

Figure 33: Predicted classes and proportion of offspring in matings between heterozygous overos. Darker shading emphasizes the homozygous overo class.

The percentage of lethal white foals observed by breeders might be less than 25% because the overo parents may include more than one genetic type of overo, a non-lethal combination. The percentage of lethal whites to be expected by overo breeders would then be much more complicated to predict, related both to the relative frequency of the overo types and to the tendency of the mare owners to use a stallion of the same overo pattern as the mare. Another factor that could reduce the actual frequency of lethal white would be an association between the homozygous condition and early pregnancy loss.

Genetic relationship between tobiano and overo

Analysis of test cross data shows the genetic relationship of tobiano and overo. If tobiano and overo are alleles of one gene, tobiano/overo (tovero) stallions bred to QH and TB mares would produce only tobiano or overo foals, no solids and no toveros. However, studbook data from such horses invariably show four pattern classes of foals, in approximately equal numbers. Thus, separate, unlinked genes produce the tobiano and overo patterns (Figure 34).

Genetic contribution from solid dams	Genetic contribution from tovero stallion (*OoTOto*)			
	O TO	*o to*	*o TO*	*O to*
o to	*Oo TOto* tovero	*oo toto* solid	*oo Toto* tobiano	*OO toto* overo
Offspring proportion	25%	25%	25%	25%

Figure 34: Offspring of a tovero stallion (medicine hat) in test cross matings show independence of genes for tobiano and overo since the four phenotypic classes occur in a 1 : 1 : 1 : 1 distribution.

Tovero stallions may be nearly white (medicine hat) showing an additive effect of the traits' gene actions. Tobiano and overo interaction apparently can also produce a white horse. Among Paint horses it is probably more common for a white foal to be the product of overo parents and to die within hours or days of birth. However, viable white Paints occur rarely. Breeding records show at least some of them to have foals of all four pattern classes.

Defects associated with overo

Occasional Paint foals from overo breedings are born as blue-eyed whites, or nearly white with a few colored spots about the muzzle or a few colored hairs about the ears or tail (Hultgren 1982, McCabe *et al.* 1990, Schneider & Leipold 1978, Smith 1977, Vonderfecht *et al.* 1983). Nearly all show symptoms of intestinal discomfort within a few hours of birth, similar to a foal with retained meconium. Neither medication nor surgery is successful to overcome the blockage. These foals cannot pass food through the digestive tract, due either to lack of nerve cells (myenteric and submucosal neuronal plexuses) that control the peristaltic muscle actions of the gut or, more rarely, to missing intestinal tract sections. The disease has been called lethal white foal syndrome (LWFS) or ileocolonic aganglionosis. Nearly all affected foals are the products of overo × overo mating, but a rare few exceptions are reported. The association of a neurological defect with a conspicuous and unusual pigment pattern is encountered in other species and may relate to the fact that during embryological development both nerve cells (ganglia) and pigment cells (melanocytes) are migratory cells that originate from the same neural crest area.

An obvious model to explain the inheritance of the recurring problem is that lethal white foals are the expression of the homozygous condition for overo. In single gene dose, the only manifestation of an abnormal gene is an attractive white spotting pattern. How does this proposal explain the occurrence of lethal white foals when only one parent is an overo? Possibly the solid parent is genetically an overo, but has insufficient white spotting to be recognized as an overo for APHA registration purposes. Alternatively, the second overo gene could arise under the same circumstances that produce cropouts. Under this model, lethal white foals would also be possible (extremely rarely) as offspring of QH matings.

Deafness, only rarely identified in horses, may also be occasionally associated with overo, especially with horses that are mostly white. Much more work is needed to substantiate the anecdotal reports of deafness in overos, to determine the incidence of hearing loss and whether it is associated with a specific pattern.

Other diseases and genes associated with neural crest-derived cells

A lethal intestinal defect similar to that of lethal white Paint foals has been reported in Clydesdale foals (Dyke *et al.* 1990). The foals, aged 4–9 months, had a history of lethargy and abdominal distention. The intestinal defect was characterized as megacolon and microscopic examination of the affected gut showed an absence of ganglia. This Clydesdale problem is different from LWFS in that the foals were not white and not neonates. However, the sabino pattern is prominent in Clydesdales, so a connection of the intestinal defect to an "overo" spotting pattern is a distinct possibility. Dark-eyed white horses are found in a few horse breeds. In these breeds, white is transmitted as a dominant gene, found only in heterozygotes (Pulos & Hutt 1969).

Mouse mutant studies are also useful resources to understand the genetics of overo spotting and associated defects in the horse. At least five unlinked genes in the mouse affect pigment and nerve precursor cells embryonically derived from the neural crest and can be considered as candidates for homologous loci to overo. The mouse genes are Splotch (*Sp*) on chromosome 1, lethal spotting (*ls*) on chromosome 2, microphthalmia (*mi*) on chromosome 6, piebald (*s*) on chromosome 14 and Dominant megacolon (*Dom*) on chromosome 15 (Greene 1981, Lane & Liu 1984, Silvers 1979). The proposal that more than one gene may produce "overo" spotting of the horse is clearly supported by the mouse model.

Hirschsprung disease (aganglionic megacolon), a human genetic disease that maps to chromosome 10 (in a region homologous to the mouse chromosome 14 region of the piebald gene), provides another model for understanding the LWFS (Angrist *et al.* 1993, Lyonnet *et al.* 1993). Hirschsprung disease differs from the horse problem: it does not involve a pigment disorder and can often be surgically corrected. Although long suspected to be an inherited problem, the genetics of Hirschsprung disease was variously thought to be recessive, polygenic or dominant until recent mapping studies defined a single major gene locus and confirmed a dominant trait transmission.

Deafness is associated with Waardenburg's syndrome, another human disease with a distinctive pigment distribution (white forelock). One form of Waardenburg's syndrome has been mapped to the human homolog on chromosome 2 of the mouse *PAX3* gene (Baldwin *et al.* 1992). The mouse spotting pattern trait Splotch is assigned to the *PAX3* locus. Another form of Waardenburg's syndrome is on human chromosome 3, at a site homologous to the mouse microphthalmia (*mi*) gene, associated with white spotting and hearing loss (Tassabehji *et al.* 1994). Deafness is also associated with white color or spotting patterns in cats, Dalmatians and dogs homozygous for the merle pattern.

Genetic linkage group for overo

No genetic linkage for overo has been reported, but few family studies have been directed toward looking for such relationships. Studbook study shows that overo and tobiano are not linked.

Comparing tobiano and overo genes

Some people see tobianos as white horses with dark patches and overos with the opposite scheme. In minimally marked tobianos the white body markings generally appear as vertical stripes, but in overos the white patches more often show a horizontal spreading. Exceptions to these generalizations occur often, particularly when multiple spotting genes occur in one horse. Several bits of information including patterns of parents and patterns of offspring should be used for assignment of genotype.

Overo and tobiano are dominant genes, but the patterns have genetic differences that tend to blur the basic similarity. Tobiano inheritance is predictably regular and homozygotes occur. Overo can occur from matings of solid color horses (particularly in specific QH bloodlines), is not known to be true breeding (homozygous), and is associated with lethal effects.

Frequency and origin of cropout overos

About 500 cropouts have recently been registered annually by the APHA compared with an AQHA annual registration of just over 100,000. These figures suggest roughly one in 200–300 offspring of QH parents has white markings sufficient to be registered as an overo by the APHA. No *simple* model is sufficient to explain the origin of the cropouts in Quarter Horses and the subsequent transmission of the overo trait in Paint breeding programs.

One possibility for the origin of cropouts is as mutation events. In this model a cropout arises infrequently as a mutation in a particular gene whose altered state produces white spotting patterns. The mutation rate is higher than is traditionally seen for most genes (a mutation "hot spot"). Among potential mechanisms of mutation are insertion of long or short interspersed elements, retroviral elements or DNA instabilities in regions of gene duplication (e.g. Brilliant *et al.* 1991, Chabot *et al.* 1988, Copeland *et al.* 1983a, Copeland *et al.* 1983b, Geissler *et al.* 1988). The mutated gene will produce overo markings (perhaps by failure of pigment cell migration) and is passed on as a dominant gene, lethal when homozygous.

Another possibility avoids the mutation hypothesis by assuming that the overo phenotype is the result of a particular allelic combination rarely occurring in Quarter Horses. The logical extension of this scheme to expain the breeding results of overo in Paint horse programs is not clear.

The inheritance of some spotting patterns in mice is also complex, and those situations may provide important clues to help us understand overo genetics. Extensive molecular genetics research will be necessary to provide the exact explanation for cropouts, but fortunately examples of similar types of gene action in other organisms define starting points. Without a molecular marker for overo it is difficult to predict the likelihood of producing a cropout from any particular pair of QH parents.

Predicting the likelihood of an overo foal by an overo stallion

Among traditional advertising statements for overo Paint stallions, a "colored" foal percentage from breedings to Paint mares is often given. The typical percentage seems to be about 90%, but any data percentages that include offspring from Paint mares confound the true evaluation of the stallion's contribution to the foals. Some of the dams might have been homozygous tobianos from which even a solid colored stallion would be able to chalk up credit for a spotted foal. The pattern breeding potential of an overo stallion can be determined by the proportion of overo offspring among his foals from solid mares. Current models predict that that percentage will typically approximate 50% or 75%, reflecting how many dominant spotting genes he has. With such test cross data a mare owner can predict the real likelihood of a particular mare producing a spotted foal by the stallion being considered.

Breeding Stock overos

Solid color offspring of matings between overos or an overo and a solid parent can be registered with the APHA as Breeding Stock. Since overo patterns generally are dominantly inherited traits rather than recessive, a prudent assumption for Breeding Stock horses will be that they are solid because they have failed to receive an overo gene. Exceptions and minimally marked overos that do not meet the APHA definition certainly may occur. Overo pattern probably can be suppressed by other genes, for example genes that produce horses without leg or facial white markings. Until we know the DNA sequence of overo and can look for it with a diagnostic test, we have no reliable way to determine which minimally marked horses have an overo gene.

The best way to produce an overo foal using a Breeding Stock horse is to use an overo mate. Breeding Stock horses may be a source of valuable conformation- or performance-related genes useful in a breeding program, but they should not be overestimated as a reliable source to contribute overo pattern genes to an overo breeding program.

CHAPTER 7
Leopard (appaloosa) spotting

This complex of spotting and diffuse roaning patterns of variable extent is relatively symmetrically dispersed over the hips, down and forward (Figure 35). Mottled skin, white sclera and striped hooves are also characteristic. The white pattern of an individual horse may increase in extent to about age five, but at least some pattern manifestations are present at birth, such as mottled skin. The patterns can occur with any basic coat color and with the other major spotting genes (tobiano and overo). The basic color may be slightly diluted or darkened along with the spotting effects. Often the spots have a different texture from the surrounding coat, especially noticeable in winter coat.

The origin of this pattern predates written history, but its spread worldwide is attributed to the influence of the Spanish horse. In North America the pattern is most closely associated with the Appaloosa, a breed tracing to horses of the Nez Perce tribe from the Palouse River region, now Idaho.

Trait inheritance and gene symbol

Probably a single major gene acting as an incomplete dominant is responsible for all the patterns, with pattern diversity attributed to modifying genes (Sponenberg 1982, Sponenberg *et al.* 1990). Homozygotes for the spotting gene usually have a greater extent of white than heterozygotes, and are popularly called few-spot leopards. The gene symbol is *LP*.

Few-spot leopard	*LPLP*
Leopard, blanket, varnish roan, snowflake, frosted	*LPlp*
Not patterned	*lplp*

Although theories invoking multiple genes have been proposed by other authorities to account for the variation in patterns, Sponenberg and colleagues (1990) convincingly demonstrated that a single major gene with

Figure 35: One pattern type produced by the leopard spotting (appaloosa) (*LPlp*) is illustrated by this black Appaloosa stallion with a blanket of distinctive spotting over the croup and hips.

minor gene modifiers was adequate to explain studbook and family data from a variety of breeds.

Some of the patterns of difusely sprinkled white hairs (varnish roan (Figure 36), snowflake and frosted) could be confused with effects of gray or roan, but careful study of the colors of the horse and its parents should be able to resolve the issue. For example, except for the occurrence of scars, the distribution of white hairs in a roan is relatively uniform over the body. The varnish roan, produced by leopard spotting, has a mottled roaning pattern with darker areas over facial bones, hips, knees and hocks.

Figure 36: An Appaloosa mare illustrating the varnish roan variation of leopard spotting (*LP*), occasionally confused with the roan (*RN*) allele.

The gray gene epistatically suppresses the coat pattern characteristics of a mature Appaloosa horse, so many breeders avoid gray horses in their color breeding programs.

Linkage

Definition of gene linkage within a family or breed could help clarify whether a single gene produces the variety of patterns seen, but no linkage studies have been reported.

Breeds

This spotting gene is found worldwide in pony, draft and light horse breeds such as Appaloosa, Miniature Horse, Pony of the Americas, Mongolian pony, Knabstrup and Noriker. It is not immediately obvious how a Spanish breed origin for the trait can include the Mongolian pony and Miniature Horse, but the Spanish element in the other breeds is clear. In the Noriker, selection has strongly favored the leopard spotting type, while in the Appaloosa many horses have the blanket or varnished roan phenotype. Occasional examples of spotted foals occur in breeds that have selected against the trait (e.g. Quarter Horses and Paints), suggesting a minimally marked type may be difficult to distinguish or *LP* is masked by interaction with other genes.

CHAPTER 8
Putting it all together: color genotypes and phenotypes

All horses will have a pair of alleles for each of the 11 color and pattern genes listed here. This chart shows how you can assign genotypes by phenotype observation. For each box with more than one choice and for all boxes with a " ∼," assignment of genotype requires information from parents or offspring.

	Color genes										
	White	Gray	Black/ red		Dilution			Pattern			
Coat color	*W*	*G*	*E*	*A*	*C*	*D*	*Z*	*TO*	*O*	*LP*	*RN*
White	*Ww*	∼	∼	∼	∼	∼	∼	∼	∼	∼	∼
Gray	*ww*	*GG* *Gg*	∼	∼	∼	∼	∼	∼	∼	∼	∼
Bay	*ww*	*gg*	*EE* *Ee*	*AA* *Aa*	*CC*	*dd*	*zz*	*toto*	*oo*	*lplp*	*rnrn*
Bay tobiano	*ww*	*gg*	*EE* *Ee*	*AA* *Aa*	*CC*	*dd*	*zz*	*TOto* *TOTO*	*oo*	*lplp*	*rnrn*
Bay varnish roan	*ww*	*gg*	*EE* *Ee*	*AA* *Aa*	*CC*	*dd*	*zz*	*toto*	*oo*	*LPlp*	*rnrn*
Black	*ww*	*gg*	*EE* *Ee*	*aa*	*CC*	*dd*	*zz*	*toto*	*oo*	*lplp*	*rnrn*
Black roan	*ww*	*gg*	*EE* *Ee*	*aa*	*CC*	*dd*	*zz*	*toto*	*oo*	*lplp*	*RNrn*
Chestnut	*ww*	*gg*	*ee*	∼	*CC*	*dd*	*zz*	*toto*	*oo*	*lplp*	*rnrn*
Palomino	*ww*	*gg*	*ee*	∼	$C^{cr}C$	*dd*	*zz*	*toto*	*oo*	*lplp*	*rnrn*
Palomino overo	*ww*	*gg*	*ee*	∼	$C^{cr}C$	*dd*	*zz*	*toto*	*Oo*	*lplp*	*rnrn*
Buckskin tobiano/overo	*ww*	*gg*	*EE* *Ee*	*AA* *Aa*	$C^{cr}C$	*dd*	*zz*	*TOTO* *TOto*	*Oo*	*lplp*	*rnrn*
Red few-spot leopard	*ww*	*gg*	*ee*	∼	*CC*	*dd*	*zz*	*toto*	*oo*	*LPLP*	*rnrn*
Red dun	*ww*	*gg*	*ee*	∼	*CC*	*DD* *Dd*	*zz*	*toto*	*oo*	*lplp*	*rnrn*
Cremello	*ww*	*gg*	∼	∼	$C^{cr}C^{cr}$	∼	∼	∼	∼	∼	∼
Grulla	*ww*	*gg*	*EE* *Ee*	*aa*	*CC*	*DD* *Dd*	*zz*	*toto*	*oo*	*lplp*	*rnrn*
Silver dapple	*ww*	*gg*	*EE* *Ee*	*aa*	*CC*	*dd*	*ZZ* *Zz*	*toto*	*oo*	*lplp*	*rnrn*

CHAPTER 9
White markings

White markings on the head and legs are prominent characteristics of domestic animals. Usually these markings have underlying pink skin. In contrast to the regular markings of white-patterned wild species, white markings in domestic animals are often asymmetrical. In a study of foxes raised for pelts, the Soviet geneticist Belyaev (1979) found white markings appeared spontaneously in breeding lines selected for domestic "dog-like" behavior. These experiments suggested that markings can be physiologically associated with characteristics that distinguish domestic from wild animals.

The presence of markings in domestic horses is one of the conspicuous features that distinguishes them from their closest wild relatives, Przewalski's horses. Markings are so common in domestic horses of most breeds that it is often the *unmarked* horse (Figure 37) that is singled out for comment.

Markings can help make horses readily identifiable as distinct individuals and make interesting subjects for folklore traditions. Markings have been used to provide guidelines for purchase ("One white foot buy him, two white feet try him, three white feet look about him, four white feet do without him") or prognostication of fate (e.g. lightning is more likely to strike a horse with an irregular blaze than one unmarked or symmetrically marked), but such folklore about markings has not been substantiated by scientific study. Some horse breeds have been selected to be without markings (Cleveland Bay, Friesian) and others have distinctive and extensive markings (Shire, Clydesdale). For Quarter Horses registration is barred for horses having extended markings on the head, legs or body. The Arabian horse is not characterized by any defined markings pattern; indeed, the breed is known to have a variety of possible patterns, although occasional breeders have chosen to emphasize certain subsets of the array.

Is it possible to predict the markings of an unborn foal based on the markings of the parents? Intuitively it is obvious that although markings traits are inherited, the genetics must be complicated since generally the only information one can find is in the form of hearsay advice.

What causes white markings?

Pigments are produced by specialized cells called melanocytes. These cells originate during embryonic development from the neural crest (backbone)

Figure 37: A dark bay Quarter Horse stallion with no white markings.

region and migrate outward. Lack of pigmentation can be due to failure of melanocyte migration, or inhibition or failure of melanin synthesis. For head and leg markings of horses, it is likely that a failure of melanocyte migration produces the areas of traditional markings defined by pink skin and white hair, although other explanations are possible. White markings without pink skin are usually only found as small or narrow areas on the face or as edges on larger spots. The dark skin under those markings is probably a result of a limited migration of melanocytes from adjacent pigmented areas. These later migrating cells may not infiltrate the hair follicles so the hair in those areas remains white, or white mixed with color.

From studies in mice, it is likely that markings in horses could be due to genetic variation at a gene homologous to piebald. Other genes may be involved as well. In mice, markings on feet, face, tail and around the umbilicus occur in animals homozygous for a recessive allele at piebald. Occasionally heterozygotes may have markings, depending on the presence of alleles at other spotting loci. Genetic analysis becomes complex since markings produced by piebald can appear to be inherited either as a dominant or a recessive trait, depending on genetic background.

Regardless of what names are given to the common markings genes in horses, some are likely to interact with “major” spotting genes including tobiano, overo and appaloosa to modify those patterns. The modifications could include extending white patterns as well as suppressing them to the equivalent of a conventional marking.

Non-genetic influences

No two horses with any kind of white markings have exactly the same patterns. This diversity is important for the easy identification of horses as distinctive individuals. While this diversity of markings probably reflects the great variety of alleles available to influence melanocyte migration, it also could be produced by non-genetic influences. For any trait being analyzed for genetic variation, it is important to understand the influence of non-genetic factors on the appearance of the trait. White spots can be due to destruction of melanocytes in localized areas as a consequence of injury. More subtle influences are also at work. A clue to the involvement of non-genetic influences on markings is the asymmetry of leg markings on most horses (Figure 38). All four legs of an individual horse have exactly the same genetic makeup, but one leg may be without markings, another with a coronet marking, and the other two with white half-way to the body—or any of an endless variety of possible combinations.

Classical genetic research studies that seek to understand non-genetic influences may use identical twins. Since naturally occurring twins in horses are fraternal (non-identical), identical twins in horses can only be obtained as a consequence of embryo splitting research. Allen and Pashen (1984) split embryos into approximately equal halves and implanted each in a different mare. When the foals were born, the twins were of like sex and color, but the white markings were not the same. For example, in one twin pair, one colt had socks on all four legs and the other colt had socks on only the left fore

Figure 38: A dark chestnut Arabian mare demonstrates asymmetrical leg markings.

and hind legs. The non-genetic components that influenced markings differences were not identified, but nonetheless the research dramatically illustrated that any consideration of the genetics of markings must include the understanding that not all the observed variation between individuals is due to a difference in genes.

Polygenic inheritance

Traits that are influenced by non-genetic factors typically involve the interaction of several to many genes, each with a small effect, acting together additively. Such traits may be called polygenic or multifactorial and a clear understanding of their involvement in any particular trait usually includes mathematical models and computer analysis.

Heritability (symbolized h^2) is used to provide information about the amount of variation in a trait that is due to genetic differences. Heritability estimates are traditionally provided for quantitative traits such as egg or milk production or weaning weights of calves, for which an environmental component (e.g. nutrition) can be as important as genes. Heritability values range from 0 to 1 and are obtained for a specific population (breed or subset). The higher the value, the greater the accuracy of prediction of genetic components by a consideration of the phenotype of a given animal within the studied population. Since h^2 values are estimates, they can vary between experiments. Their validity is strengthened through the aggregation of several reports whose estimates are within comparable range.

Woolf (1989) analyzed facial markings inheritance in Arabian horses. For the data set he selected 22 sires with at least 100 foals and 7343 dam–foal pairs. He divided the face into areas, but he did not consider the amount of white in each area. His study showed that several pairs of genes, each with a small effect, act additively to control the number of areas of facial white. He found no evidence that each area of the face was controlled by a single gene. Thus the genetics of white facial markings appears to be controlled in a polygenic or multifactorial manner. The heritability was estimated to be 0.69, which means that about two-thirds of the variation observed in white facial markings is produced by genetic differences. A statistical profile for expected facial markings of foals from specific matings could be generated by reference to tables in his 1989 paper. For example, 66% of foals from parents without facial white also had no facial white, 25% were marked with a star and 9% had facial white in addition to the star area. For parents with facial white in all five scored areas (star, strip, snip, upper lip and lower lip), 54% of foals were similarly marked to the parents and 46% had fewer areas with facial white than the parents.

The heritability estimate for white markings in Arabians cannot be assumed to apply for all horses. Similar studies are needed for other breeds, particularly by breeders of Quarter Horses and Morgans, who must comply with regulations restricting registration to horses with markings not exceeding defined boundaries.

Influence of coat color and sex

Woolf (1990) extended the markings study to include leg markings and the influence of major genes, including coat color and sex. He found that the number of areas showing white facial markings is correlated with the number of legs showing markings, suggesting that they are the product of the same genetic mechanisms. The heritability estimate of white markings on legs was 0.68, and the heritability of white markings considering scores for both leg and facial white was 0.77. His data corroborated the influence of color on markings shown in other breeds (chestnuts have more white-marked areas than bays) and also the influence of sex (males have more extensive white markings than females) (Dreux 1966). In a later paper Woolf (1991) provided data showing a greater extent of white markings in heterozygous (*Ee*) than homozygous (*EE*) bays in Arabians. Thus markings are influenced by a complex of factors, including sex, a polygenic markings system, the *E* locus and non-genetic factors.

These studies show that a breeding program, at least in Arabians, that selected breeding stock with respect to white markings could with reasonable assurance of success ultimately produce foals with either no white or extensive white markings. Certainly any such program would be a long term proposition, and it would be difficult to select for other traits while concentrating on markings.

CHAPTER 10
Parentage testing of horses

Genetic testing validates horse pedigrees for breed registry authorities, sales companies and race tracks. In the USA alone over 50 horse registries have some form of genetic testing requirement. Worldwide nearly 40 laboratories perform a similar panel of genetic tests using blood type markers. Combining the efforts of all laboratories, genetic marker records include a million horses, conservatively estimated. Current breeding stock tested in the last decade probably constitute more than half the records, dramatically illustrating the interest in horse pedigree verification.

The genetic markers used for parentage testing are inherited according to principles of Mendelian genetics. A standard test battery for horses typically consists of 15 systems of blood group and protein markers. Of paramount concern to breed registries and owners are the accuracy of results, the cost, and the effectiveness of the tests to detect incorrect parentage. The blood typing tests performed by qualified laboratories worldwide have proven to be highly accurate, repeatable, and affordable to breeders of purebred horses. The efficacy of the standard test to detect an incorrectly identified sire or dam when the other parent is recorded accurately is about 97–99%, depending upon breed and systems used (Bowling 1985). Detection of pedigree errors arising from outcrossing to another breed or switching one horse for another approaches 100%. DNA testing techniques now available extend the number of genetic markers that can be used and provide an alternative testing method if needed.

Blood group testing

Horse blood groups are analogous to the human red blood cell (RBC) types such as "O negative" and "A positive." Human blood factors are probably most familiar as inherited differences that are identified to select the appropriate blood donor for transfusion, but they are also determined as part of the process to prevent the maternal–fetal blood incompatibility problem of newborns known as Rh disease, to answer parentage questions, or to identify criminal suspects from blood stains left at a crime scene. Horse blood group factor testing has the same applications, but parentage verification is the most common use.

Reagents for horse blood group testing (serological testing)

Blood grouping reagents (antisera) are made for each species (human, bovine, equine and so on) by carefully planned immunization routines. Antisera to detect horse blood group factors are generated by injecting a recipient horse with a small amount of blood from a selected donor horse. The donor–recipient pairs are chosen to differ by as few known factors as possible, preferably by a single factor, so that the natural immune process of the recipient will raise a monospecific blood grouping antiserum. The antiserum potentially provides a reagent that can be used to test thousands of horses, since very small amounts are needed for each test and properly stored reagents (frozen or freeze-dried) can last for years.

Standard nomenclature for horse blood group factors

As of August 1994, the Horse Standing Committee of the International Society for Animal Genetics (ISAG) distinguishes 34 blood group factors distributed in seven systems (A, C, D, K, P, Q, U). Factors within each system are coded by letters that are assigned in order of their recognition by the Committee. Each system has from two to 25 possible types designated by letter combinations (Table 5).

System	Factors	Recognized alleles
A	a b c d e f g	A^{a} A^{adf} A^{adg} A^{abdf} A^{abdg} A^{b} A^{bc} A^{bce} A^{c} A^{ce} A^{e} A^{-}
C	a	C^{a} C^{-}
D	a b c d e f g h i k l m n o p q r	D^{adl} D^{adlnr} D^{adlr} D^{bcmq} D^{cefgmq} $D^{cegimnq}$ D^{cfgkm} D^{cfmqr} D^{cgm} D^{cgmp} D^{cgmq} D^{cgmqr} D^{cgmr} D^{deklr} D^{deloq} D^{delq} D^{dfklr} D^{dghmp} D^{dghmq} D^{dghmqr} D^{dkl} D^{dlnq} D^{dlnqr} D^{dlqr} D^{q} (D^{-})
K	a	K^{a} K^{-}
P	a b c d	P^{a} P^{ac} P^{acd} P^{ad} P^{b} P^{bd} P^{d} P^{-}
Q	a b c	Q^{abc} Q^{ac} Q^{a} Q^{b} Q^{c} Q^{-}
U	a	U^{a} U^{-}

Table 5: Seven loci of horse blood group markers, the factors detected by antisera and the alleles (phenogroups) that are recognized by the International Society for Animal Genetics (ISAG).

Extensive family studies have shown that the combinations of factors in each system are transmitted as a group (phenogroup) and inherited as co-dominant traits (Bowling & Williams 1991, Sandberg 1973, Scott 1978, Stormont & Suzuki 1964). Individual laboratories may have additional factors in their reagent batteries that await completion of requirements for international recognition by ISAG: 1) at least two laboratories must raise antisera for the factor; 2) the laboratories must demonstrate concordant results on a panel of 40 horses in an ISAG Horse Comparison Test; 3) the laboratories must exchange antisera to confirm concordant results of the antisera on at least 200 more horses; and 4) the laboratories must publish family data demonstrating segregation relationships with previously recognized factors.

Serological tests to identify blood group factors

In the analytical test, RBCs from each horse are diluted in saline (salt) solution and mixed with aliquots of a battery of extensively tested, functionally monospecific antisera (reagents). The presence of a factor is recognized either by clumping of RBCs (agglutination), or by lysis of RBCs (hemolysis) in the presence of complement. Rabbit serum is used as a source of the complement enzyme cascade that recognizes RBCs complexed with antibody and subsequently ruptures the cell membrane, releasing hemoglobin. Nonreactive samples are scored as negative for the selected factor. Blood group frequencies vary between breeds (e.g. Bowling & Clark 1985).

Testing of protein polymorphisms (biochemical polymorphisms)

Electrophoresis is a technique that uses an electrical current to separate a mixture of molecules embedded in a supporting medium (starch, agarose or acrylamide gel). When applied to blood proteins, electrophoresis can reveal genetic differences between animals. The distance that a protein migrates under standard conditions is a genetic trait related to the protein's electrical charge, size and shape, all of which are conditioned by its amino acid sequence, in turn determined by nucleotide base sequence in DNA.

Serum samples and RBC lysates are complex mixtures of proteins sorted out by electrophoresis and then visualized by protein-specific dyes or histochemical stains. The resulting band patterns represent protein products of genes.

Currently, the ISAG recognizes 16 systems of protein variants (Table 6), for which at least two member laboratories produced compatible results in comparative testing. These systems are chosen to be effective for parentage testing. TF and PI are highly polymorphic (with many alleles).

For the most part, the molecular basis of the observed differences has not been determined. This does not prevent the use of protein markers in parentage testing. Reliable discrimination of genetic variants is a paramount criterion and has been repeatedly verified for blood groups and proteins through analysis of family data and international comparison and standardization tests.

Blood samples from members of family groups demonstrate that codominant alleles produce the band pattern variation for a specific protein. Genotypes are directly determined from phenotypes. Genetic markers and frequencies vary between breeds (e.g. Bowling & Clark 1985). Seldom will one breed show all the variants described, but certain variants are common in nearly all breeds. Data for several breeds are provided in Table 7 for the serum protein transferrin, a highly polymorphic trait. The commonly found markers are TF-D and TF-F_2, in all breeds from ponies to draft horses.

System	Locus symbol	Recognized alleles
A1B glycoprotein	*A1B*	*F K S*
Albumin	*ALB*	*A B I*
Acid phosphatase	*AP*	*F S*
Carbonic anhydrase	*CA*	*E F I L O S*
Catalase	*CAT*	*F S*
NADH-diaphorase	*DIA*	*F S*
Carboxylesterase	*ES*	*F G H I L M (N) O R S*
Group-specific component	*GC*	*F S*
Glucose phosphate isomerase	*GPI*	*F I L S*
Hemoglobin-α	*HBA*	*A AII BI BII (C) N V*
Peptidase A	*PEPA*	*F S*
6-Phosphogluconate dehydrogenase	*PGD*	*D F S*
Phosphoglucomutase	*PGM*	*F S V*
Protease inhibitor	*PI*	*F G H I J K L* L_2 *N O P Q R S T U V W Z*
Plasminogen	*PLG*	*1 2*
Transferrin	*TF*	*D* D_2 *E* F_1 F_2 F_3 *G* H_1 H_2 *J M O R*

Table 6: Genetic systems of blood biochemical polymorphisms used for parentage verification. Markers in parentheses await family analysis and validation through international comparison tests. Additional rare alleles have been reported in several of these systems, but generally appear to be restricted to a single breed or small families within a breed. Methods for performing these tests are described in various publications (Bengtsson & Sandberg 1973, Bowling *et al.* 1988, Bowling *et al.* 1990, Braend 1970, Juneja *et al.* 1978, Kelly *et al.* 1971, Pollitt & Bell 1980, Sandberg 1968, Sandberg 1974b, Scott 1970, Weitkamp *et al.* 1983, Yut & Weitkamp 1979).

An interesting but unanswered question is why horses have so much detectable genetic variation for this iron-binding blood protein gene. One possibility is that the detection methods are highly favorable for easily identifying genetic variation in transferrin, but not so effective for identifying variation in proteins at other loci, such as albumin, that have only two or three alleles. Alternatively, the extensive allelic array for transferrin may not be an artifact of technique but may have functional characteristics that favor selection of heterozygotes (perhaps as viability or breeding fitness traits), thus maintaining a high level of polymorphism in the population or breed. Until we have DNA sequence data to compare a significant number of horses at the nucleotide level, we will be unable to know whether those blood protein systems with many alleles are truly more genetically variable than loci with fewer detected alleles.

Rare variants are characteristic of highly polymorphic systems. Some rare variants within a breed may be a consequence of undetected crossbreeding in previous generations. Others may trace to breed founders whose lines

Breeds	Transferrin variants													
	[A]	D	D_2	E	F_1	F_2	F_3	G	H_1	H_2	J	M	O	R
AN	+	0.29	Ø	+	+	0.28	Ø	Ø	Ø	0.29	+	Ø	0.05	0.08
AR	Ø	0.26	Ø	+	+	0.50	+	+	+	0.12	Ø	Ø	0.11	+
IC	Ø	0.20	Ø	Ø	Ø	0.46	Ø	Ø	+	0.03	Ø	0.01	0.16	0.14
LI	Ø	0.04	Ø	0.11	Ø	0.20	Ø	Ø	Ø	0.15	Ø	Ø	0.42	0.08
MH	+	0.31	+	+	+	0.54	0.01	0.01	+	0.06	+	0.02	0.02	0.03
MI	Ø	0.13	0.01	0.01	+	0.41	+	0.03	+	0.03	Ø	0.04	0.16	0.18
QH	+	0.26	+	+	0.17	0.34	+	+	+	0.07	Ø	+	0.05	0.10
SH	Ø	0.31	0.07	Ø	Ø	0.47	Ø	Ø	Ø	Ø	Ø	Ø	0.13	0.03
ST	Ø	0.23	Ø	+	Ø	0.54	+	+	Ø	+	Ø	Ø	0.06	0.17
TB	Ø	0.31	Ø	+	0.34	0.19	+	+	+	0.02	Ø	+	0.06	0.09

Table 7: Frequencies for genetic variants of the serum protein transferrin in ten breeds. Polymorphic systems such as this are highly effective for recognizing incorrect parentage. The 14 alleles are coded alphabetically. The breeds are: Andalusian (AN); Arabian (AR); Icelandic (IC); Lippizaner (LI); Morgan Horse (MH); Miniature Horse (MI); Quarter Horse (QH); Shire (SH); Standardbred (ST) and Thoroughbred (TB). A + indicates the variant was found, but at a frequency less than 0.01. An Ø indicates the variant was not found among the tested animals. Brackets indicate a variant not currently included in the list of internationally recognized variants. The calculated effectiveness (using Jamieson 1965) of this single locus to detect incorrect parentage ranges from 36% of cases in Morgans to 55% in Miniatures and Quarter Horses.

have not been among the predominant contributors to the contemporary gene pool. At this stage in many breeds' evolution, it is probably not possible to determine which type of origin accurately describes most rare variants.

Lymphocyte testing (histocompatibility markers)

A complex, genetically controlled system of tissue antigens called the major histocompatibility complex (MHC) is highly effective for causing immune-mediated rejection of tissue transplanted between unrelated animals. Human lymphocyte antigen (HLA) testing was developed in the 1970s and 1980s and has proven to be spectacularly useful for matching tissue transplantation recipients with compatible donors. In addition, HLA has been used as a highly effective system for parentage assignment, and can be used to predict susceptibility to certain autoimmune diseases.

The development of equine lymphocyte antigen (ELA) tests for horses was eagerly anticipated in the early 1980s, particularly for application to parentage testing and disease association studies. Two linked loci of class I ELA systems were defined for horses (*ELA-A* and *ELA-B*) (Bernoco *et al.* 1987a, Bernoco *et al.* 1987b) and assigned to the long arm of chromosome 20 (Ansari *et al.* 1988, Mäkinen *et al.* 1989). The tests can be applied to solve parentage cases (Bailey 1984), but have not proven to be as feasible as using blood group, protein or DNA polymorphisms.

A susceptibility to viral-based sarcoid tumors is associated with ELA determinants (Lazary *et al.* 1985).

DNA testing of horses

Horse breed registries have made extensive use of blood group and protein polymorphism testing as part of the process to ensure pedigree integrity. DNA-based tests for parentage verification are becoming increasingly available and will probably replace conventional testing. DNA testing has great popular appeal as well as scientific merit. DNA markers can be determined from tissues other than fresh blood, including hair or carcass material.

Options for DNA testing techniques

DNA tests for animal parentage must have the qualities of conventional testing, e.g. accurate, effective, inexpensive, rapid, with ready transfer of results between laboratories, and will have the attractive features that they may be highly automated and will not be limited to fresh blood. DNA techniques and markers that could be used include multilocus fingerprints (minisatellites) ("genetic profiling"), restriction fragment length polymorphisms (RFLPs), biallelic systems (single nucleotide polymorphisms, SNPs), mitochondrial sequence polymorphisms (mtDNA) and microsatellites, also known as short tandem repeats (STRs) or simple length repeats (SLRs). Sorting out the best choice for the animal breeding industry is currently under active investigation.

Fingerprinting techniques are discussed at length in the scientific and popular press, but are not under consideration for breed registry parentage testing programs. Although fingerprinting could be applied as a solution to individual cases (Bernoco & Byrns 1991, Georges *et al.* 1988, Troyer *et al.* 1989), for animal industry programs it requires too much DNA, takes too much time to produce and analyze results, and is too difficult to apply recorded results to another case, let alone to cases in another laboratory. RFLPs share nearly the same problems as fingerprinting for an animal industry application (requiring too much DNA and processing time), although they, too, could be useful for individual cases and may have applications for genetic mapping (Harbitz *et al.* 1990, Kay *et al.* 1987a, Kay *et al.* 1987b, Rando *et al.* 1986). Mitochondrial polymorphisms have the potential to be powerful tools for sorting out maternity, but cannot answer questions about paternity. Certain biallelic systems are attractive, particularly in not requiring electrophoresis to detect variants (Nikiforov *et al.* 1994), but discussions among animal testing laboratories currently favor microsatellites as the option that has the best spectrum of properties needed for animal parentage verification programs.

Microsatellites

Microsatellites are composed of simple tandem repeats of nucleotides primarily in non-coding gene sequences (DNA regions not used as templates for protein synthesis). For example, one of the first published horse microsatellites (HTG6) consists of a string of repeats of the two DNA bases T and G (Ellegren *et al.* 1992). DNA sequences flanking the repeat region uniquely define HTG6 somewhere in the horse genome (chromosome assignment presently undefined, as for most of the blood group and protein systems). Among horses studied so far, the number of TG repeats at HTG6 has varied from four to 26. The number of repeats is inherited as a codominant (allelic) trait. A uniform nomenclature for reporting phenotypes has not yet been standardized for horses, but agreement is expected shortly following discussion of international comparison testing results.

In case you are still hungry for more "alphabet soup" jargon, remember that microsatellites are also known as STRs (short tandem repeats). They are also examples of STSs (sequence tagged sites) and VNTRs (variable number of tandem repeats). For a slight measure of relief from acronyms of molecular biology and genetics, I prefer the name microsatellite in this discussion.

Genetic variation in microsatellites is detected using **PCR (polymerase chain reaction)** and electrophoresis. The starting material is some source of DNA (blood and hair roots are the commonly available tissues). DNA primers bracketing the microsatellite region direct amplification of a specific chromosomal segment. Starting with as little material as a single gene copy, the PCR process duplicates that DNA sequence several millionfold in just a few hours. Assay of tandem repeat numbers (alleles) for each horse is achieved by analysis of fragment size after electrophoresis of PCR products. Alleles can be determined by computer-assisted laser detection of fluorescent dyes tagging the amplified products or by less complex technology such as silver staining or isotopic labelling. The allelic information determined for each sample is easily stored in computer databases and shared between laboratories.

As seen with blood groups and protein markers, some variants are common in most breeds. Frequencies for variants of the dinucleotide microsatellite VHL20 (van Haeringen *et al.* 1994) are provided in Table 8. Note that the variant designated as M is relatively frequent in four breeds, but infrequent in Arabians. In contrast, variant L, while common in four breeds, is rare (but still present) in Miniatures. Thoroughbreds have less genetic diversity than the other four breeds shown.

Thousands of other examples of dinucleotide repeats like HTG6 and VHL20 (e.g. Breen *et al.* 1994, Ewen Matthews 1994, Guérin *et al.* 1994), as well as mono-, tri-, tetra- and pentanucleotide repeats, are expected to be found in the horse genome. In addition to parentage testing, microsatellites are useful for gene mapping, which may lead to new selection tools to control genetic diseases and to improve performance traits.

Analysis of variation in mitochondrial sequences

Analysis of mitochondrial DNA sequences has been a powerful tool in human forensics to confirm maternal relationships, including those of skeletal remains in unmarked graves. Mitochondrial sequences from bones believed to be from bodies of the last Tsar of Russia and his family have

	VHL20 variants									
Breeds	I	J	K	L	M	N	O	P	Q	R
AR	0.11	0.01	Ø	0.33	0.04	0.22	Ø	0.01	Ø	0.26
MI	0.16	0.12	Ø	0.01	0.27	0.29	0.08	0.01	0.02	0.03
QH	0.26	+	+	0.26	0.28	0.14	0.01	0.02	+	0.03
ST	0.15	0.02	Ø	0.24	0.27	0.14	0.01	0.03	0.02	0.12
TB	0.33	Ø	Ø	0.22	0.29	0.16	Ø	Ø	Ø	Ø

Table 8: Frequencies of variants for microsatellite VHL20 in five horse breeds. The ten alleles are coded alphabetically from smallest (I) to largest (R) according to the size of the PCR fragment. The breeds are: Arabian (AR); Miniature Horse (MI); Quarter Horse (QH); Standardbred (ST) and Thoroughbred (TB). A + indicates the variant was found, but at a frequency less than 0.01. An Ø indicates the variant was not found among the tested animals. The efficacy of this system to detect a falsely assigned parent ranges from 0.48 in TBs to 0.63 in STs.

recently been matched to living maternal relatives (including Britain's Prince Philip) several generations removed from Tsar Nicholas and his wife (Gill *et al.* 1994).

Identifying mitochondrial sequence polymorphisms is a complicated process, and may be too expensive to apply routinely to verify maternity in horses. However, recent publications show that polymorphisms in horse mitochondrial sequences exist and could be applied to the solution of special cases (Gerlach *et al.* 1994, Ishida *et al.* 1994, Xu & Årnason 1994).

Using DNA systems for routine parentage analysis of horses

The genetic variation within microsatellite loci provides a method for parentage verification based on new DNA technology that requires only minimal changes in the efficient routines developed for blood typing. Although the technology is complex, parentage testing by microsatellites is simple in comparison with the traditional blood group and protein polymorphism testing because only a single technique is used for all systems.

The task now at hand is to identify horse microsatellite genes that are highly polymorphic, allowing test panels to be composed of a relatively small number of systems. A dilemma facing the animal industry and the laboratories is how to achieve the switch to the new technology. The markers detected with the conventional test have no corresponding DNA-based analytical test. If a parentage case is to be analyzed with DNA markers, horses previously tested for blood group and protein markers will need to be tested for the DNA markers. This testing could be accomplished from stored blood, but if a stored sample is unavailable, a new sample (blood or hair roots) will need to be submitted.

During the several decades of its use, genetic testing for pedigree verification has served the animal industry well. The testing laboratories and the breed registries for which they provide services are enthusiastic about the possibilities for incorporating DNA testing into parentage verification programs. Some registries are already in the process of converting to tests based on DNA markers, and others expect to follow shortly.

Parentage problems in horses

Registries may require genetic testing only for horses involved in artificial insemination with fresh or frozen semen, or embryo transfer, but many authorities have extended their requirements to natural breeding as well. Pedigree errors occur infrequently in most breeds, but just a few examples of genetic marker solutions for potentially difficult problems convince horse registry officials and owners of the need for parentage verification programs.

Efficacy of tests to detect falsely assigned parentage

The efficacy of a single system or locus depends on the number of alleles, their frequencies and whether the genotypes can be directly determined from the phenotypes (Rendel & Gahne 1961). The most effective are loci having five or more alleles with appreciable frequencies and without null (recessive) alleles. In the conventional blood typing test the most effective loci have probability of exclusion (PE) values for recognizing a falsely assigned random parent close to 0.6. Two such loci would detect about 84% of false parentage cases and three loci about 94% of cases. A standard test battery that has been used internationally for nearly a decade includes 15 systems (A, C, D, K, P, Q and U of blood groups; ALB, A1B, ES, GC, HBA, PGD, PI and TF of blood proteins) and has a calculated effectiveness of 97–99% depending on breed for the detection of randomly assigned false parentage. Additional loci can be added so that the efficacy approaches 100%.

Sample case results and analysis

A foal qualifies as an offspring of its parents if the factors detected in its tissue sample can be accounted for by applying Mendel's laws to the factors found in tissue samples of the parents. The **Mendelian law of dominance** excludes a foal as a mating product when it possesses a factor not present in at least one of the parents. The **Mendelian law of segregation** excludes a parent that fails to share a genetic marker with a foal assigned to it. These are the same laws we applied to inheritance of coat color variants in previous chapters, even though we did not provide the name of the laws in that context.

Table 9 provides an example of phenotypes for horses involved in a hypothetical paternity case. Note that none of the five horses has exactly the same combination of letters and thus the individuals are clearly genetically distinguishable.

	Blood group	Serum proteins				Microsatellite
	D	ALB	TF	PI	GC	VHL20
D (dam)	dkl	AB	DH	LU	F	LM
O (offspring)	dghmqr/dkl	AB	DR	LN	F	MN
S1 (stallion 1)	bcmq/dkl	B	DO	N	FS	LM
S2 (stallion 2)	dghmrq/dlnqr	A	DR	NU	F	N
S3 (stallion 3)	bcmq/cegimnq	B	HO	FG	F	LQ

Table 9: Phenotypes for six selected systems of blood genetic markers in a hypothetical parentage case in which a mare was known to have been exposed to three stallions. Genotypes can be directly inferred from phenotypes. Blood group alleles (phenogroups) in heterozygotes are separated by a slash ("/"). Heterozygous phenotypes for serum proteins and the microsatellite are reported as two-letter combinations. Single-letter phenotypes are presumed to represent homozygotes.

Comparison of the foal's type with that of the dam shows that the sire of this foal must possess Ddghmqr, TF-R, PI-N and VHL20-N. **S1** lacks Ddghmqr, TF-R and VHL20-N. The foal is excluded as an offspring of **D** and **S1** (Mendelian law of dominance). **S3** fails to share a genetic marker with the offspring and is excluded as a sire of this foal in the D, TF, PI and VHL20 systems (Mendelian law of segregation), without consideration of the dam. **S2** possesses the required markers and qualifies as a sire of this foal. Breed registries consider verification of parentage is a sufficient test for pedigree validation; statistical likelihood calculations are seldom required.

If the dam could not be tested for the case in Table 9, neither **S1** nor **S2** could be excluded as a sire of this foal based on these data. Tests of additional genetic systems could be applied until evidence for exclusion of one of the sires was obtained.

Sire exclusion in routine parentage verification

The widespread use of castration as an efficient selection tool limits the number of breeding males. A mare is usually bred to a single stallion by natural service or by artificial insemination. If genetic testing identifies a sire exclusion for a foal, the alternative breeding males are usually limited and easily defined based on the situation at the farm or stable where the mare was boarded 11–12 months prior to the birth of the foal.

Before considering other sires, it may be prudent to verify that the correct samples were submitted for testing. Samples may be taken from the wrong horse by persons, delegated by the owner, that may not clearly understand which horses need to be tested. Occasionally tissue sampling kits may be switched between two horses sampled at the same time. In such cases the exclusion result is obviously correct. The foal owner, who had no reason to question the foal's parentage, will initially presume the problem is with the laboratory's testing procedures. It may take some time to sort out the muddle, assign the genetic types to the correct horses and finally verify the offspring's parentage.

Dam exclusion in routine parentage verification

As mentioned in the sire exclusion section, before trying to sort out a possible pedigree error it is important to eliminate a sampling error. When the genetic type is verified to be correctly assigned to the designated (identified) horse, then procedures to correct the pedigree can be productively tackled.

Foals are customarily registered at an early age so maternal assignment is highly likely to be correct. Mares may switch foals, but the circumstances that lead to the situation are very rare. For example, blood typing excluded maternity in two similar cases in which a bay foal was said to have been born to chestnut parents, incompatible by genetic principles of coat color. Blood marker analysis in these cases confirmed the proposal that two mares foaling on the same night in a small paddock had switched foals. Exclusion of maternity more commonly is a consequence of animals becoming switched after weaning, after sales (transport agent mixes up lots), after arrival at a new facility (new owner/manager/trainer assumes identity of stock without reference to markings on registration certificates), or of animals being similarly marked or not marked at all so that it is difficult to tell them apart.

Mating exclusion in routine parentage verification

The foal is excluded as an offspring of a mating when each parent individually qualifies as a sire or dam but the foal has genetic markers not found in the parental combination. Three possibilities need to be considered: 1) the dam may be correct, but not the sire; 2) the sire may be correct, but not the dam; or 3) neither of the proposed parents is correct. Usually, the standard test battery narrows down the three scenarios to one so that owners can review their records and identify one or a few parentage options to test. If the owner cannot provide alternative suggestions, additional tests might be applied to help direct the search to another sire or dam or both.

Multiple sires

Another application of genetic testing is to eliminate all but one sire for an offspring when its dam was known to have been exposed to more than one breeding male. A recurring example of such a situation involves a two-year-old filly with an unanticipated foal. As a yearling she may have been pastured with two (or more!) yearling colts, one of which is likely to be the sire of the foal, but mating was never observed. From our laboratory records, we find that genetic analysis excludes paternity (as assigned by the owner) for 25% of offspring born to two-year-old fillies, a striking contrast to the 0.5% exclusion frequency in random testing of offspring from dams unselected for age. Often the multiple males in a parentage case are related, as sire and son or as sons of the same sire. If systems in a standard battery of tests are genetically linked, the greater likelihood of shared haplotypes (linkage phases) among relatives would reduce the effectiveness of the tests to solve the paternity question. Linked systems occur in the present battery of blood typing tests, but the highly polymorphic systems are independent from each other. With the extended test battery (including DNA-based tests) available today, it should be possible to resolve every multiple sire case.

Single-parent cases

Occasional cases require comparison of genetic markers from a single parent and its alleged offspring. One parent may be unavailable because the horse is dead or its whereabouts unknown, or the owner refuses to supply a tissue sample. The accuracy of a conclusion that excludes parentage is not affected, but single-parent tests are less effective than two-parent comparisons for detecting a genetic incompatibility. Highly polymorphic systems with many detectable variants (e.g. D, TF and PI) prove to be particularly useful in single-parent verifications. Additional systems may need to be added to the standard test battery to achieve an acceptable level of exclusion probability.

Resolution of controversial cases

Most parentage exclusion cases are resolved amicably. Seldom do cases go to court; if adjudication is necessary the breed registry hearing process is usually sufficient. Mismanagement is probably more often the cause of exclusion than fraud. Since the qualifying pedigree may be judged less valuable than that on which the breeding contract or purchase was based, most court cases are civil suits for settlement of damages.

Forensic cases

Crime or fraud cases involving horses are rarely presented for testing, but that situation may change since testing of DNA markers does not have such stringent requirements for sample and sample condition as do blood group and protein polymorphism tests. Genetic testing may be applied to verify whether the genetic type of a sample used for drug or antibody titer tests matches a second sample taken from the identified animal in the presence of witnesses. Theft cases may be solved by genetic comparisons that exclude one proposed pedigree. Occasionally the unavailability of one set of parents (perhaps by deliberate destruction of animals) confounds the resolution. Derived genetic types of dead animals, using their known offspring, has been used for genetic exclusion for the proposed pedigree.

Identification

Systems that are highly effective for parentage determination also provide a largely unique genetic profile for identification of an individual horse. How likely is it for two horses to have the same genetic markers? The answer will depend on breed and genetic systems used, but we will focus on the possibilities for Quarter Horses. Using ten loci of microsatellites (HTG4, HTG6, HTG7, HTG10, HMS1, HMS6, HMS7, AHT4, AHT5 and VHL20) *the most common type* is calculated to occur with a frequency of 8.1×10^{-8} or about 1 in 12 million unrelated horses. Clearly two tissue samples with identical markers using such a genetic characterization panel would be highly likely to have been obtained from the same horse or identical twins. So far identical twins have not been demonstrated to occur naturally in horses, although they have been produced experimentally with embryo splitting experiments.

To substitute one horse for another (a "ringer") without detection initially requires that external marks of signalment (sex, color, age and markings) are identical. Clearly it is virtually impossible to match two horses for external signalment *and* genetic markers. The detection of ringers can be easily achieved once genetic marker profiles have been established for a group of horses.

Parentage inclusion

Knowing that the conventional battery of blood typing tests is 97–99% effective to detect incorrect parentage, does that mean that the pedigree of a particular horse that has been parentage verified has a 1–3% chance to be incorrect? The answer is no, the possibility of incorrectly assigned parentage when no exclusion has been identified is usually very, very remote. The tedious calculations must be performed on a case-by-case basis so probabilities are seldom determined, except in disputed cases.

Stated another way, if we have a parentage match (parentage inclusion), what is the probability that the verified pedigree is correct? As an example, consider the following case. A Standardbred mare was sent to a stud farm for breeding to a particular stallion. Subsequent routine parentage verification testing excluded the foal as an offspring of the selected stallion. Considering the five other stallions present on the stud farm, four were also excluded and one stallion qualified. The mare owner sued the stud farm to obtain a breeding certificate so that the qualifying stallion could be listed on the foal's registration papers. The stallion owner did not contest the maternity of the foal, but refused to accept the genetic marker data as substantial evidence for the foal's paternity. The probability that a random Standardbred stallion qualifies as the sire, given that the dam is accepted as correct, was calculated to be one in 344,828, considerably exceeding the number of Standardbred stallions available in North America, let alone the state where the breeding occurred. In this case it is clearly highly likely that the Standardbred stallion that qualified and was on the farm at the same time as the mare is the true sire of this foal. The case was settled out of court with a property settlement between the stallion and mare owner.

What can I learn from my horse's genetic marker report?

A genetic marker report (Figure 39) provides four types of information. One section gives the **sample identification** information provided with the blood sample to the testing laboratory. This includes name, registration number, color, sex, year of birth, breed, sire name, sire registration number, dam name and dam registration number. Also in this section is a reference number assigned by the laboratory to the sample, and the date of the test.

```
Name: GEORGE 112345
Sex: S  Breed:  TB        YOB: 90        Color: C
Tested:   5/16/91   VGL Case: TXXXXXX
Sire:   Magnum 15488
Dam: Fleur 97362

Blood Group Factors
        A: adf/b         C: a     D: bcmq/dkl
        K: a     P: ad   Q: abc  U: -      T: TV
Electrophoretic Systems
        ALB: AB        TF: DR    XK: K     ES: I
        HBA: BI/BII    GC: F     PI: LN    PGD: FS
Factors recognized:   Aabcdefg  Ca   Dabcdefghiklmnopqr
Ka   Pabcd   Qabc   Ua    T:TV   AL:A,B,I;TF:D,E,F1,F2,F3,G,H1,
H2,J,M,O,R,*;XK:F,K,S;ES:F,G,H,I,L,N,S,*;PGD:D,F,S,*;
HB: BI,BII,A,AII,*;GC:F,S:PI:F,G,I,K,L,L2,N,S,T,U,W,*

Veterinary Genetics Laboratory, University of California
Davis, California  95616-8744       USA
```

Figure 39: Genetic marker report for blood type factors.

Another section provides the **blood group factors** of the named horse detected with blood group reagents. No blood group factor is confined to a single breed. We know of no evidence that predicts any particular performance traits based on detected factors. Information about blood group factors may be useful to identify compatible whole blood donors and to manage the anemia problem in foals known as neonatal isoerythrolysis (NI). Taken together, blood grouping tests potentially recognize several hundred thousand distinctive genetic profiles.

A third section of the blood type record provides information for the named horse on genetic markers known as **protein variants**, detected by electrophoresis. In horses this information has no known medical or performance significance, but is extremely useful for parentage questions since a tremendous array of genetic types can be detected. The eight systems usually detected in a standard test using red blood cells and serum are ALB (albumin), ES (esterase), GC (group-specific component), HBA (hemoglobin-α), PGD (6-phosphogluconate dehydrogenase), PI (protease inhibitor), TF (transferrin) and A1B (A-1-B-glycoprotein, recent system name change from XK). Recognizable differences in each system are assigned letter designations using an internationally accepted nomenclature. The hundreds of thousands of protein variant combinations potentially recognizable are used to establish individual identity and to verify parentage.

A final section of the blood type record lists all the **markers recognized** by the testing techniques applied. Markers with internationally accepted nomenclature are listed by letter, but if the laboratory can detect additional markers for which an international nomenclature has not been defined, this is designated by an asterisk. The information listing markers recognized is important if the record needs to be interpreted by a second laboratory.

The format for a DNA marker report will be similar to that for blood group and protein variants, but the nomenclature has not yet been standardized.

CHAPTER 11
Medical genetics

A **congenital abnormality** is a defect of structure or function evident at birth, not necessarily caused by defective genes. An environmental influence (a nutritional deficiency or excess, an inhaled or ingested chemical, drug or toxin) that interferes with normal developmental processes may cause the observed defects. The incidence of congenital defects in foals (e.g. parrot mouth, lax or contracted tendons, patent urachus, heart defects) is estimated to be 3–4% (Rossdale 1972). Congenital problems may directly relate to foal death. To prevent recurrence of a congenital problem, it is important to know its basis. Unfortunately many defects, including those listed above, have no established cause.

Inherited diseases may be congenital, but most of the known genetic diseases of horses do not become evident for months or even years. This chapter will focus on inherited defects in horses, including the kinds of evidence required to reach a definitive conclusion as to whether a condition has a genetic basis. A discussion of carrier testing and selective breeding schemes will help breeders cope with the rare but ever-present potential of genetic problems.

Genetic diseases can be broadly divided into three groups:

- Chromosomal problems (discussed in Chapter 12).
- Simple (single gene) Mendelian traits.
- Polygenic traits.

The first two categories are ultimately the easiest to document and understand, but are rarely seen. Polygenic traits, produced by the interaction of several genes, are more likely to be the source of commonly observed defects such as those of conformation. This chapter will discuss examples among several breeds of horses that have good or reasonably good evidence to be single gene or polygenic diseases. (A few additional genetic problems are discussed in Chapter 19.) Also included in this chapter is a discussion of the medical aspects of blood group related problems.

Breeders looking for information on a certain potential genetic problem may be disappointed to find no discussion here. In the same way that molecular techniques have transformed the field of human genetics, horse breeders can anticipate that research in the next decade should provide a wealth of new information about medical genetics of horses so that definitive information can be provided to help breeders raise healthy and useful horses.

Identification of Mendelian gene defects

Usually the first evidence of a genetic problem is the occurrence in repeated matings of foals that exhibit similar defects of structure or function. It is unlikely in the extreme that all offspring of a given mating or an entire foal crop would be affected by a genetic disease, so such a situation would be presumptive evidence of a non-genetic (environmental) problem. An important clue that a problem could be genetic is its association with a particular breed. Breed association is more significant of genetic disease than familial or herd association because animals related by pedigree or ownership are likely to have encountered the same environmental conditions.

Dominantly inherited mutants may have extreme effects on the organism. Animals with such genes seldom survive to breed, although some dominantly inherited diseases such as HYPP (discussed below) provide an exception. Most identified genetic anomalies are inherited as recessive genes, meaning that the parents are unaffected. Recessive genes "hidden" in heterozygotes by the dominant normal gene may be more widespread in a breed than intuitively recognized. *Deleterious mutations that occur in purebred breeds are usually chance "hitch-hikers" in highly successful breeding lines, otherwise the homozygous genotypes that produce the problem conditions would be so rarely encountered as to be overlooked or written-off as problems with unknown causes.*

Pedigree studies of affected animals may provide tentative evidence for trait inheritance, but since most animals within a breed tend to be somewhat related, pedigree relationships alone cannot be used to establish the inheritance of a defect. Clinical and morphological study may relate the abnormalities to a proven genetic disease in another species. Verification of the inheritance of the trait in horses requires data documenting the number of affected and unaffected foals produced from selected informative matings. To test genetic hypotheses for problems in which the affected animals do not die before reproductive age, it is possible to design matings between them, and between affected and normal animals. If the defect is produced by a lethal gene, matings to collect genetic data can only be designed between animals that have sired or produced an affected foal (presumptive heterozygotes) and many offspring will be needed for a meaningful test of the genetic hypothesis.

Once the necessary research to prove the single gene basis of a defect has been completed, the breeder's work has just begun. If a laboratory carrier test is available, for example by analysis of a blood sample, breeders may choose to breed only from non-carriers or, alternatively, to use carriers, but select breeding stock in subsequent generations that are free of the problem gene. If no laboratory carrier test is available, the avoidance of breeding with carriers is an imperfectly lofty goal. While every effort may be made to avoid breeding from carriers, a large number of carriers will by chance remain undetected. *Clearly, a laboratory carrier test is a highly desirable goal for the breeding management of any genetic disease.*

Musculoskeletal defects

Hyperkalemic periodic paralysis (HYPP)

Sporadic episodes of generalized muscle tremors, stiffness and paralysis not associated with exercise, accompanied by elevated levels of serum potassium, are caused by a dominantly inherited genetic defect in Quarter Horses and breeds using QHs as breeding stock (Cox 1985, Steiss & Naylor 1986). The abnormality has been identified to be in the gene controlling the muscle sodium channel protein (Rudolph *et al.* 1992). Affected animals usually appear to be heavily muscled, a trait valued by QH breeders. The bulkiness of HYPP positive horses may be a sign of muscle pathology, not increased fiber numbers. For most horses with the defective gene, the disease can be controlled by regular exercise and a diet low in potassium. The effects of the mutant gene are not usually seen in foals, except that foals with two copies of the gene (homozygotes) show signs of respiratory distress as neonates and may not survive.

The gene defect in HYPP has been identified to be a single nucleotide change that results in the replacement of the amino acid phenylalanine with leucine in the sodium channel gene of muscle protein. This change produces abnormal muscle physiology. A gene test based on analysis of the DNA sequence (using PCR and restriction enzyme technologies) is available for positive identification of animals with the trait.

The gene test (Figure 40) allows horse owners to make informed decisions about purchasing or breeding from horses that are candidates to have the abnormal gene.

	Normal allele	HYPP allele
DNA sequence	...ATC TTC GAC TTC...	...ATC TT*G* GAC TTC...
Amino acid sequence	...isoleucine-phenylalanine-asparagine-phenylalanine...	...isoleucine-*leucine*-asparagine-phenylalanine...
Test result code	N	H

Figure 40: HYPP is associated with a single DNA and amino acid sequence difference from the normal muscle sodium channel allele.

- N/H horses have a single copy of the mutant sequence, are candidates to have episodes of muscle paralysis and will on average transmit the trait to 50% of their foals (Spier *et al.* 1993).

- H/H horses are likely to be severely affected and have poor breeding or performance prospects.
- N/N horses do not have the abnormal DNA sequence and will not produce affected foals if mated with other N/N horses.

Most genetic diseases that concern animal breeders are inherited as recessive genes. HYPP is produced by a dominant gene and provides a contrasting case study. Its origin seems to trace to a single prominent stallion, found in the pedigrees of over 100,000 Quarter Horses. The defective muscle sodium channel generally produces only mild problems in heterozygotes. Many gene positive (heterozygous) animals appear to be asymptomatic to their owners. This is exactly as anticipated for a wide-spread dominant defect. If the effects of the mutation drastically affected the health of every carrier, the gene would be self-limiting.

Myotonia with dystrophic changes

Episodes of prolonged muscle contraction associated with voluntary muscle stimulation have been reported in young Quarter Horses (e.g. Jamison *et al.* 1987). The clinical and pathological symptoms are similar to a dominantly inherited disease in humans but no genetic studies have been done with affected horses.

Hereditary multiple exostosis (HME)

HME is a rare skeletal dysplasia consisting of multiple benign bone neoplasms detected most commonly at sites of active bone growth such as the metaphyseal regions of long bones, as well as ribs, scapula, pelvis and vertebrae. An affected Thoroughbred stallion sired both normal and affected sons and daughters when bred to normal mares, and thus demonstrated the trait is inherited as an autosomal dominant (Gardner *et al.* 1975).

Flexural limb deformities associated with complete ulnas and fibulas

A congenital condition of severely splayed legs (Figure 41) was described by Speed (1958) to be associated with a full-length ulna and fibula (Figure 42), normally reduced in horses, that improperly articulated with the carpus (knee) or tarsus (hock). Body size is disproportionately reduced due to relative shortening of the humerus and radius of fore leg and femur and tibia of hind leg (short-legged dwarfism). Affected foals are born of normal parents.

Genetic studies of the condition were performed by Hermans (1970) with Shetland Ponies. The bone defects were proposed to be inherited as an autosomal recessive trait, and this model was verified by experimental matings. Flexural limb deformities associated with complete ulnas and fibulas have also been reported in a Welsh pony (Shamis & Auer 1985) and the figures in this section document the occurrence of this problem in Miniature Horses.

Early ancestors of the horse, several million years ago, had complete ulnas and fibulas, but this characteristic has been lost in the course of modern horse evolution. Reappearance of ancestral characteristics in individual

Figure 41: Severely splayed hind legs ("cow hocks") in a Miniature Horse filly born to normal parents are characteristic of a fibular malformation described by Speed (1958) in Shetland Ponies.

members of the species are called **atavisms**. They provide evidence that genetic information for embryological development may be retained, even if a particular structure has disappeared from a species (Hall 1995). Polydactyly is another atavism occasionally seen in horses. A change in an embryological developmental pattern could be due to genetic factors but a non-genetic teratogenic insult during gestation could also produce such characteristics. Breeding trials are an important tool for understanding whether the basis of an atavism is genetic or environmental.

Lateral patellar luxation

In a limited breeding trial resulting in 49 foals, Hermans and colleagues (1987) showed that a congenital defect of the stifle joint in Shetland ponies is likely to be inherited as a simple recessive trait. Affected foals were born from normal parents. Test cross foals were produced in a 1:1 ratio of affected to normal, as expected for a recessive. The genetic hypothesis for patellar luxation must be weighed in each specific case against the possibility of the problem being due to traumatic injury or other environmental factors.

Occipital-atlanto-axial malformation (OAAM)

Neurologic impairment ranging in severity from ataxia to tetraplegia in Arabian horses has been reported as a consequence of spinal cord compression and injury. Affected foals are often unable to stand and nurse, although in

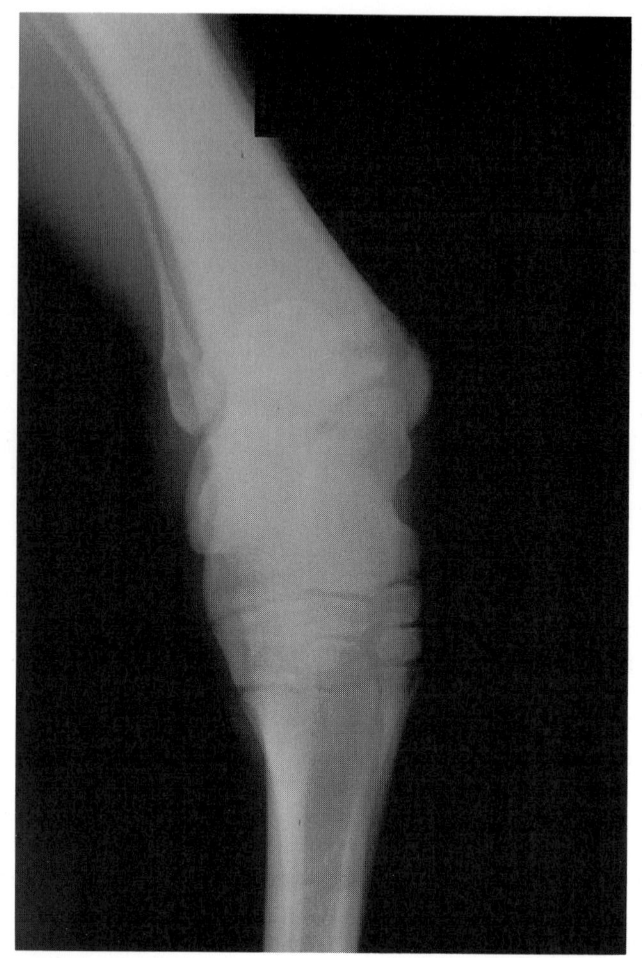

Figure 42: Radiograph of hock region of splay-legged filly pictured in Figure 41. Notice the fibula (small, thin bone at upper left, parallel to the heavier tibia) articulating with the tarsus.

some cases the symptoms may take a few weeks to become obvious. The spinal cord lesions are associated with malformation and fusion of the cervical vertebrae to the skull (Leipold *et al.* 1974, Mayhew *et al.* 1978, Watson & Mayhew 1986). The trait is suggested to be a lethal autosomal recessive trait, but breeding trials have not been reported.

Brain and other neurologic defects

Narcolepsy

Excessive sleepiness, depression and collapse in two closely related Miniature Horse foals were diagnosed as narcolepsy (Lunn *et al.* 1993). Between episodes the foals were bright and alert. Symptoms could be precipitated by handling. Narcolepsy in humans is inherited as a dominant trait with incomplete penetrance, and in dogs as an autosomal recessive. No studies have identified the mode of inheritance in horses.

Equine degenerative myeloencephalopathy

A degenerative syndrome of ataxia associated with central nervous system lesions reported in several breeds (Standardbred, Arabian, Appaloosa, Thoroughbred and Paso Fino) and equine species in zoos (Grant's zebra and Przewalski's horse) has been associated with a familial sensitivity to low levels of vitamin E (Mayhew *et al.* 1987). The disease may be controlled by vitamin E supplementation. No specific mode of inheritance was proposed.

A disorder such as this, which is likely to be a consequence of both a particular genetic and an environmental (nutritional) scenario, can be quite difficult to work out for a large animal model because many offspring are needed to provide valid conclusions. Horses will benefit from candidate gene approaches that seek to identify homologous small animal or human models where the pathophysiology has been clearly defined. Owners and their veterinarians may then have the information they need to know if the problem in genetically susceptible animals can be controlled through appropriate nutrition.

Myoclonus

Quick involuntary muscle contraction in response to tactile, visual or auditory stimuli in two young Peruvian Pasos was shown to be associated with a deficiency of inhibitory glycine receptors in the spinal cord (Gundlach *et al.* 1993). In other species, including cattle (Hereford) and mutant spastic mice, the disease is inherited as an autosomal recessive trait, although no breeding trials have confirmed a mode of inheritance in horses.

Lethal white foal syndrome (LWFS) or ileocolonic aganglionosis

Occasional Paint foals born as blue-eyed whites, or nearly white with a few colored skin spots about the muzzle or a few colored hairs about the ears or tail, show symptoms of intestinal discomfort within a few hours of birth, similar to a foal with retained meconium (Figure 43). All such foals reported in the veterinary literature have had at least one overo parent (Hultgren 1982, McCabe *et al.* 1990, Schneider & Leipold 1978, Smith 1977, Vonderfecht *et al.* 1983). Neither medication nor surgery successfully overcomes the blockage. These foals cannot pass food through the digestive tract, either due to lack of nerve cells (myenteric and submucosal neuronal plexuses) that control the peristaltic muscle actions of the gut or, more rarely, to missing intestinal tract sections. An obvious model to explain the inheritance of the problem is that lethal white foals are the expression of the homozygous condition for overo. For a more detailed discussion refer to Chapter 6 (Overo).

Figure 43: Minimally marked overo mare with her newborn white foal by an overo stallion, presented for clinical evaluation of the foal's symptoms of intestinal discomfort. Histopathological study showed lack of nerve cells (ganglia) in the intestinal tract, a lethal trait.

Several years ago it was proposed that lethal white foals were the result of blood incompatibility. Mare owners were advised to prevent the problem by injecting pregnant mares with blood from the stallion. Such a recommendation would be tantamount to guaranteeing NI problems for the foal and it is to be hoped that few owners followed the recommendation. Subsequent research showed that NI problems are not detected in lethal white foals (Smith 1977).

Cerebellar disease

Cerebellar disease in the horse is characterized by ataxia (gait abnormality), lack of balance equilibrium and head tremors. Cerebellar disease has been documented among young animals of both sexes, in Arabians (DeBowes *et al.* 1987, Palmer *et al.* 1973), Oldenbergs (Koch & Fischer 1951) and Gotland Ponies (Björck *et al.* 1973).

The clinical and pathologic features seen in each breed are similar but the diseases may be caused by different genes. Pathological features associated with the behavior patterns in Arabians include loss of Purkinje cells and neurons from the granular layer of the cerebellum (cerebellar abiotrophy). The first symptoms of ataxia usually do not occur until the foal is a few weeks old. At that time, the owner may start to notice that the foal may crash into fences and fall over backwards. The foal also shows exaggerated action, particularly of the fore legs, and at rest will show a wide-based stance. Affected foals progressively become more ataxic, but not necessarily to a debilitating stage that necessitates euthanasia, although they would clearly be unsafe as riding horses. Gotland Ponies show pathological features in the molecular and granular layer of the cerebellum, with slight Purkinje cell degradation. In Oldenbergs, degenerating areas in the cerebellum lead to rapid foal death. The conditions appear to be inherited as autosomal recessive traits.

Dilute lethal

Although not yet described in the veterinary literature, Arabian owners recognize a foal lethal associated with dystocia (difficult foaling), failure to stand and nurse, neurologic problems including intermittent joint rigidity and rapid eye movements, and dilute coat color (Figure 44). Both colts and

Figure 44: An Arabian colt with dilute, pewter colored coat, joint rigidity and neurologic problems. He was never able to stand to nurse despite intensive care.

fillies are reported. Another name for this problem is lavender or pink foals. If the dilute coat color characteristic is overlooked, often the foals are diagnosed as having neonatal maladjustment syndrome, associated with brain anoxia during parturition. However, postmortem examination has identified a brain lesion (an anomalous choroid plexus) that may be the source of the neurologic symptoms. An autosomal recessive gene is proposed to account for this problem.

Blood and immunologic defects

Hemophilia A (Factor VIII deficiency)

The inheritance of certain traits that appear only to affect males can be clearly understood in light of the transmission pattern of sex chromosomes. An example of an X-linked gene is hemophilia A, reported rarely in Thoroughbred, Quarter Horse and Standardbred colts (Archer 1961, Henninger 1988, Hutchins *et al.* 1967). A defective gene, inherited from the dam, causes failure of the blood clotting process. Symptoms of hemophilia include episodes of recurrent subcutaneous hematoma, hemarthrosis and internal hemorrhage with anemia, the last commonly being the cause of death. Carrier females show no obvious effects since a normal gene copy is available on their other X chromosome to serve as a template to make the functional protein (Factor VIII) required for blood clotting. Half the daughters of dams that have produced affected colts will be unaffected carriers like their dams (see Figure 9 in Chapter 1 for a Punnett square diagram). Since an affected colt would be unlikely to be used for breeding, fillies with hemophilia (which must have received a defective gene from each parent) will not occur. In humans with this disease, clotting factor replacement therapy allows affected males to lead reasonably normal lives. Such therapy would probably be too costly for colts who are necessarily neither prime breeding nor performance candidates.

Agammaglobulinemia

A Thoroughbred colt described by McGuire and colleagues (1976) was generally healthy until about five months, when recurrent episodes of fever and lung congestion developed, followed by arthritis, laminitis and death at 17 months. Immunologic testing showed the colt lacked the ability to make antibodies, although his cell-mediated immune responsiveness was within normal limits. The features of the case are similar to those of X-linked agammaglobulinemia in people, but the inheritance of this problem in horses has not been studied.

Severe combined immunodeficiency disease (SCID or CID)

CID is a lethal disease of Arabian foals inherited as an autosomal recessive (Poppie & McGuire 1977). The disease can be tentatively diagnosed from a blood sample of a young foal. A white blood cell (lymphocyte) count of less than 1000 per mm^3 (count for a healthy foal is 2500–3000 per mm^3) and lack of IgM (immunoglobulin M) are presumptive evidence of CID. A postmortem examination provides the third diagnostic feature—underdeveloped thymus and lymph nodes. Lacking T- and B-lymphocytes of a competent immune system, CID foals succumb before five months of age to massive infection, primarily of the respiratory tract (McGuire *et al.* 1974).

Skillful application of laboratory tests is critically important to distinguish CID foals from foals whose immune system has been compromised by environmental (non-genetic) circumstances. For example, young foals who fail to receive adequate colostral antibodies (failure of passive transfer) may have low levels of immunoglobulins and increased susceptibility to infections.

An affected foal provides evidence of carrier status for both parents. No laboratory tests have yet been developed to recognize carriers that have never produced an affected foal. Matings to known carrier mares could provide the necessary progeny data to test a stallion. Because a mare can produce only a limited number of offspring, it is nearly impossible to progeny test a mare in her lifetime unless embryo transplantation is used to increase the number of foals. Clearly, a carrier test from tissue samples is critically important to plan breeding programs. Based on work in mice, productive research strategies may result from targetting genes of DNA repair as candidate genes.

Skin and hair defects

Epitheliogenesis imperfecta

A rare but recurrent congenital lethal skin defect of draft horse foals in Germany (von Butz & Meyer 1957) was reported to be inherited as an autosomal recessive. Affected foals had skin and hair missing on at least one limb distal to the knee or hock, and sometimes hand-sized, or larger, patches missing on the head, shoulder, back, or croup. Hooves of affected limbs lacked horn material. Both sexes were affected, and the foals usually succumbed to overwhelming infections within a few hours of birth. Recently foals of this description have been reported among American Saddlebred horses. This problem is a connective tissue gene defect.

Collagen defect

Extreme fragility of skin reported in related Quarter Horses of cutting horse bloodlines resembles an inherited genetic disease of humans, dogs and mink caused by defective collagen (Hardy *et al.* 1988, Lerner & McCracken 1978). Wounds occur easily and heal slowly (Figure 45). Additional studies in horses are necessary to prove the inheritance of the problem and identify the gene, but a tentative conclusion is that this defect is a recessive trait since the affected foals are produced by normal parents.

Epidermolysis bullosa

A lethal defect of two Belgian foals with multiple epidermal lesions and separation of the hooves from the coronary band has been described in detail (Johnson *et al.* 1988). Based on similar pathology in humans and other ani-

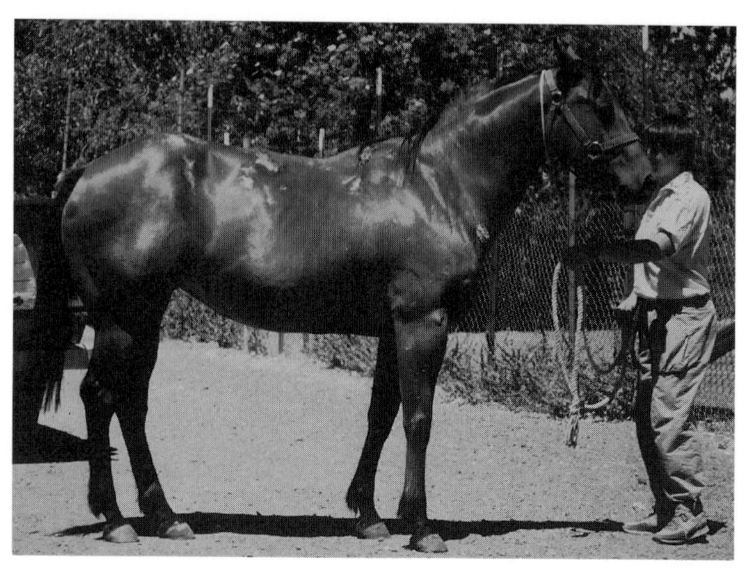

Figure 45: Skin fragility leads to spontaneous lesions and subsequent scars on the back of this Quarter Horse mare.

mals, this problem is likely to be an autosomal recessive disease involving a connective tissue gene.

Vision defects

Aniridia

A Belgian stallion with cataracts sired 143 offspring of which 65 were blind (31 males, 34 females) and 78 were normal (41 males and 37 females) (Eriksson 1955). Affected foals showed congenital bilateral complete absence of the iris. Secondary cataracts of variable onset, usually starting at two months of age, rendered most aniridic animals blind and unfit for work. The stallion's parents did not have aniridia. An autosomal dominant gene arising by spontaneous mutation was proposed to explain the defect. Normal sons and daughters of affected animals were used as breeding stock without further evidence of the iris defect. The disease is no longer considered to be a problem in Belgian draft horses.

Equine night blindness (ENB)

Equine night blindness has been recognized as a vision defect in Appaloosas (Witzel *et al.* 1977). Symptoms reported by owners include foals prone to injury during the night or older horses that stumble and fall during night riding. Visual impairment can cause the horses to be apprehensive and difficult to train. An electroretinogram can be used to confirm the clinical diagnosis. The disease appears to be similar to congenital stationary night blindness in humans. Breeding trials in horses have not been reported, but the cases reported suggest the mode of inheritance in horses as an autosomal recessive.

Genetic counseling

Breeders will want to have information about the special care required by horses affected with a genetic disorder, the potential to produce another affected horse and the long term management of their breeding programs if the affected animal was a product of their stock. The discussions are likely to be complex and filled with gray areas. Economic and emotional issues need to be discussed and decided on a case-by-case basis.

Medical prognosis for an animal affected with a genetic disorder

Lethal disorders are emotionally traumatic but special long term care requirements obviously are not an issue. Disorders that are not lethal, but debilitating, or compromise the health or use of the horse, can be major concerns for horse owners. Veterinarians are an important resource for obtaining the information about care, diet and the medical supplies that may be needed.

Breeding decisions

To avoid producing diseased foals resulting from the action of a single, dominantly inherited gene it is sufficient to avoid breeding from an affected parent. To avoid producing affected foals resulting from the effects of a single recessive gene, matings between two carriers (heterozygotes) must be avoided, but matings involving a single carrier will of course not produce an affected foal. Practically speaking, identification of carriers is an extremely difficult problem. In the absence of a specific biochemical or DNA test for the defective gene, carriers can be recognized only by having sired or produced affected foals. Many carriers will go undetected with any screening method based on *random* breedings.

A structured screening method using *planned* matings provides statistical assurance of carrier status. For example, to determine whether a young stallion carries a specific recessive lethal gene requires breeding to mares known to be carriers for that disease. If no affected foal is produced in 12 matings, then the horse is considered not to be a carrier for that gene at a 97% level of certainty. Progeny testing is slightly easier for recessive traits that are deleterious but not lethal. No affected foals in five matings with an affected mare would strongly suggest at a 97% level of certainty that the stallion is not a carrier. If a greater degree of certainty is required, more matings would be necessary. Of course, any affected foal would immediately prove that the horse being tested was a carrier for the trait under study.

To assemble the genetically appropriate breeding mares for carrier stallion screening programs is a logistically complicated proposition. In addition, these programs have the problem that while trying to reduce the frequency of undesirable genes, they will produce at least some foals that are carriers of the defective gene and thus potentially transmit it unless barred from breeding.

Carrier lists?

Some owners propose that breed registries provide lists of known carriers, particularly stallions. This would alert prospective purchasers and breeders about pedigrees in which to anticipate problems. However, sons and daughters of listed horses may be free of the undesirable gene, yet become prime suspects by association with horses named on the lists. Owners who have made the effort to understand genetics will know that 50% of the offspring of carriers will not have inherited the defective gene and they may be a source of valued genes at other loci. Other owners may not be interested in understanding the genetics of the problem and will simply be reluctant to consider any horse whose pedigree contains a listed horse, even in a distant generation.

Polygenic traits

Most traits are influenced by more than one gene. Aspects of conformation particularly fit this pattern. Progress in selection against an undesirable conformation trait can be frustratingly slow when the several individual genes affecting the trait cannot be readily identified. In addition, polygenic traits may show a **threshold effect** so that breeders may not be aware of the accumulation of problem (or desirable) genes in the selected breeding stock until a critical mass of additive genes is attained. Another complication with polygenic traits is that breeds may differ in heritability values. Even though good research data may have been compiled for one breed, the conclusions may not be directly applicable to another.

Hip dysplasia in dogs

Polygenic inheritance is so important for breeders to understand that an example from dogs will be given as one illustration of a polygenic problem. Hip dysplasia is a crippling defect of concern to dog owners and has a high incidence in many breeds. It is due to a complex set of traits, including failure of the head of the femur to fit into the socket of the hip bone and insufficient hind limb muscle mass. An imperfect fit can cause slippage and eventual osteoarthritis. Over-nutrition may cause rapid growth and be more likely to elicit the problem than conditions of slow or moderate growth. Both sexes are affected. Generally, the large breeds are most at risk. The problem may be congenital, diagnosed by radiography, but can develop any time up to 18 months or so. Eventually lameness, reluctance to exercise and hip muscle wastage become obvious. It is generally accepted that the defect is caused by a series of genes. The evidence consists of data such as: 1) the severity of the problem fits a graded series from normal through mild to severely affected, quantitated with hip scores obtained by professional evaluation of radiographs; 2) normal parents can produce affected offspring; 3) the incidence of affected offspring increases the more severe the grade of the parents' hip scores.

Despite two decades of effort, selection programs that have concentrated on reducing or eliminating breedings using severely or moderately affected animals have not reduced the percentages of affected puppies in many breeds, let alone eradicated the defect. Deleterious polygenic traits are clearly difficult to manage, but DNA technology seems to hold promise to provide direct gene tests for such problems.

Osteochondrosis in horses

Osteochondrosis is a disturbance in bone growth that can be caused by injury, fractures or malnutrition, but genetic factors have also been implicated. When bony fragments are associated with the lesion this disease is known as osteochrondritis dissecans (OCD). This is generally not a congenital disorder, but a developmental problem associated with post-natal bone growth patterns. Chronic lameness is usually the presenting problem, although the lesions do not always cause clinical signs. In a national survey of 753 Norwegian Standardbreds, the heritability of osteochondrosis in the tibiotarsal joint was 0.52 and of bony fragments in specific aspects of phalangeal joints was 0.21 (Grøndahl & Dolvik 1993). Such a study represents a tremendous effort, but the applicability of the heritability conclusions to other breeds and in other countries needs to be substantiated with additional research.

Blood groups and blood transfusions

Horses seldom need blood transfusions, but even when needed, few veterinary diagnostic laboratories are able to perform blood group testing to identify a matched donor since the blood grouping reagents are not commercially available. Veterinary laboratories are able to perform a simple hemagglutination cross-match procedure that should identify a large percentage of potential transfusion problems, even if the tests cannot identify the specificity.

Probably more than 90% of horses (breeds may be different from each other in this frequency) lack naturally occurring anti-blood group antibodies so a first unmatched whole blood transfusion is usually well tolerated, without adverse reaction. The recipient will generate antibodies against the blood group antigens of the foreign RBCs, so a second transfusion may either be ineffective due to the rapid removal of the administered RBCs or may cause an immune system crisis (anaphylaxis) that could lead to sudden death. A transfusion that did not introduce the most highly antigenic factors (particularly Aa) to a recipient that lacked such factors could probably be repeated if needed, because antibody production might only occur at low levels in response to the primary immunization (Wong *et al.* 1986).

Whenever a mare is given a whole blood transfusion, she is potentially being sensitized to blood group factors that may subsequently lead to NI problems for her foals (see later).

The naturally occurring antibodies with anti-blood group activity are usually either anti-Aa or anti-Ca. The antibodies are probably generated by immune reaction to a common environmental stimulus. Cell-surface molecules of microorganisms could be so similar to the blood group factors that antibodies to the microorganisms will cross-react with RBCs. The standard blood typing test does not identify anti-blood group activity in a horse's blood. A veterinarian should assume that any horse negative for factors Aa in the A system or Ca in the C system will have anti-Aa or anti-Ca activity. To avoid a transfusion reaction, such a horse should not receive whole blood from a donor that is positive for those factors.

Plasma transfusion

Often a blood transfusion is needed to restore fluid loss but the RBC component is not essential. In this case a plasma transfusion may fulfill the clinical requirements. Potential plasma donors can be screened to identify those without naturally occurring anti-blood group activity. Plasma can be collected and stored frozen to administer when needed.

Neonatal isoerythrolysis (NI)

A newborn foal that is healthy at birth may within 2–5 days develop signs of lethargy, elevated pulse and respiration rates and clinical evidence of anemia. Affected foals are usually from the second or later pregnancies of a mare, but first foals can (rarely) be affected. Such symptoms may indicate NI, an acute hemolytic disease of newborn foals (a maternal–fetal blood group incompatibility) caused by immunologically mediated RBC destruction. The antibody molecules that sensitize the RBCs of the foal have been passively acquired from the dam's colostrum. Recovery may be spontaneous or the disease may progress to severe anemia and death.

The most common blood group factors involved in NI are anti-Aa and anti-Qa, although other specificities have been reported (Noda & Watanabe 1975, Stormont 1975, Trommershausen-Smith *et al.* 1976b). A typical case of an Aa-sensitized mare and the available choices for stallion mates is shown in Table 10.

Not all Aa-negative or Qa-negative mares become sensitized, even though they may have produced Aa-positive or Qa-positive foals. One source of sensitization is whole blood transfusion, but this probably accounts for only a small number of cases. Mares negative for both Aa and Ca may be protected from sensitization to Aa by naturally produced anti-Ca antibodies (Bailey *et*

	A-system phenotype of stallion	A-system genotype of stallion	Foals' NI status if they receive colostrum from Aa-sensitized mare
Stallion 1	Aa-positive	*Aadf/adf*	All foals have NI
Stallion 2	Aa-positive	*Aadf/**	NI in 50% of mare's foals
Stallion 3	Aa-negative	*A*/**	No foals have NI

Table 10: Expectation of NI disease problems in foals of Aa-sensitized mare depends on genotype of stallion to which mare is bred. * denotes any A system factor other than Aa, or lack of recognized A system factor. Blood group testing may help find a compatible stallion. Breeds vary in allelic frequencies, but stallions of light horse breeds are more likely to be like Stallion 1 than Stallion 2, and less likely to be like Stallion 3.

al. 1988). At present, no simple hypothesis explains why blood group sensitization occurs in only a few percent of mares at risk for the problem.

Although it has been proposed that NI disease is associated with Arabian horses and related breeds, the practical experience of diagnostic laboratories does not support that view. Most cases seen are from Thoroughbreds and Quarter Horses, with occasional cases in Morgans and Warmbloods. Sensitized Arabian mares are rarely encountered. Also, contrary to earlier proposals, ponies are not free of the problem, demonstrated by anti-Aa cases in Miniature Horses, a very closely related breed. Certainly the frequency of sensitized mares is low in any breed and veterinarians only rarely encounter the problem.

NI is a genetic disease, but only because the blood factors involved in the immunologic reactions are inherited and the management of the disease in sensitized mares can benefit from genetic analysis. There are no data to suggest a familial susceptibility to sensitization. Horse owners need not assume that sires or dams of affected foals possess a transmissible liability that would ethically commend their removal from a breeding pool. However, no mare that has produced NI foals should ever be sold without informing the new owner of the foaling history, so that appropriate breeding or foal management practices can be applied.

NI disease in mule foals

Mule breeders are keenly aware that mares bred to jacks (male donkeys) can become sensitized to blood group factors of asses (Stormont 1975, Traub-Dargatz *et al.* 1995). Monitoring of a pregnant mare's serum for antibodies against the RBCs of the jack to which she is bred is a prudent precaution to identify mule neonates at risk for NI.

RBC donor for foal with NI

Appropriate management for a severely anemic NI foal may be to provide an RBC transfusion to alleviate the anemia and prevent death (Witham *et al.*

1984). The best blood donor for the foal is one whose cells lack the factor to which its dam is making antibodies. The sire of the foal is the worst possible donor since he has the factor to which the mare's immune system has responded. The best cell donor is the mare and she is conveniently available. The mare's RBCs should be administered to the foal in a suitable transfusion solution, but first they must be separated from the plasma that contains the antibodies reacting with the foal's cells. This transfusion is unlikely to be a match for the blood group type of the foal but it provides the foal with vital RBCs that will not be destroyed as a result of antibodies it received with colostrum from the dam. Hopefully, in a short time the dam's anti-blood group antibodies will be eliminated from the foal's system and the foal can make sufficient RBCs of its own to prevent a recurrence of the anemia crisis.

CHAPTER 12
The horse karyotype and chromosomal abnormalities

The genetic information of all horses is nearly identical and, not surprisingly, horses of all breeds have the same number, size and shape of chromosomes. We will describe the features of the normal karyotype to provide a background against which the very rare abnormalities can be understood.

Chromosomal study can detect extensive DNA alterations but not single base changes or even small deletions of nucleotide sequence within single genes. Chromosomal abnormalities associated with the addition, loss or rearrangement of a large number of genes are generally incompatible with life and are eliminated through gamete inviability or early embryonic loss before they can be observed or studied. The chromosomal defects most likely to be encountered, and then only rarely, involve the sex chromosomes and, consequently, lead to infertility. Chromosomal changes in autosomes are likely to have a much more drastic effect on viability and few examples have been found in horses. The very infrequent occurrence of chromosome abnormalities means that veterinarians may not be fully familiar with presenting symptoms.

The standard horse karyotype

The current standard karyotype for the horse was defined by Richer and colleagues (1990). It consists of 13 meta- and submetacentric (bi-armed) and 18 acrocentric pairs of autosomes, plus a large submetacentric X and a small acrocentric Y.

Four banding techniques have been used to define the unique identity of each chromosomal pair. These special procedures are used to pretreat chromosomes before staining, creating banding patterns that allow the pairs to be determined with greater certainty than with unbanded preparations (Buckland *et al.* 1976, Kopp *et al.* 1981, Romagnano & Richer 1984).

- Trypsin (a protein digesting enzyme) treatment followed by giemsa staining (**G-banding**) defines regions with a high concentration of base sequences containing adenine and thymine (Figure 46).

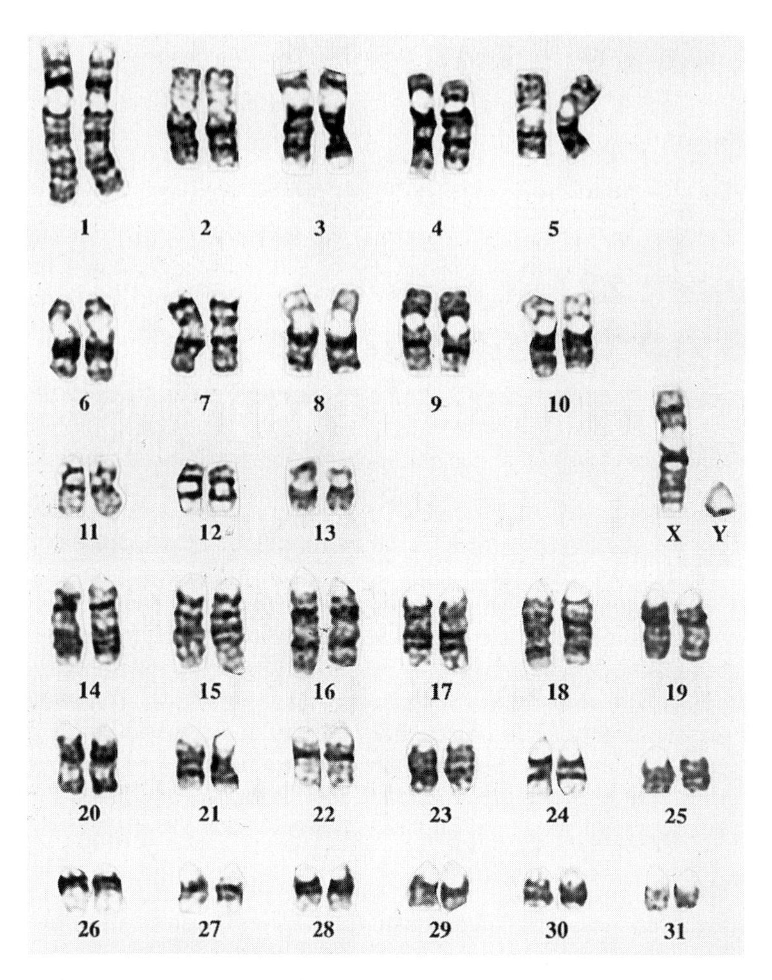

Figure 46: G-banded karyotype of lymphocyte chromosomes (of a male horse). Pairs are ordered according to descending size in three groups: the sex chromosomes on the third line at the right edge; meta- or submetacentric autosomal pairs numbered 1–13 on lines one to three; and acrocentric autosomal pairs numbered 14–31 on lines four to six (from Bowling 1992, reprinted with permission of the publisher, Elsevier).

- Barium hydroxide treatment and giemsa staining (**C-banding**) identifies heterochromatin, characteristic of most of the centromeric regions. C-banding is particularly useful for distinguishing the sex chromosomes because the Y chromosome is entirely heterochromatic, compared to the other small autosomes from which it might be difficult to distinguish by size, and the X-chromosome has an interstitial C-band on the long arms, making it distinctive from the similarly sized second largest autosome.
- BrdU (5-bromo-2′-deoxyuridine) incubation followed by giemsa staining produces reversed banding patterns (**R-banding**) compared with trypsin digest procedures.

- Silver staining identifies nucleolar organizer (NOR) regions that produce ribosomal RNA used in translating the genetic code into proteins.

Both G- and R-banding are useful for precisely identifying loss, gain or rearrangement of chromosomal segments.

Chromosomal polymorphisms not associated with disease

Normal variations (heteromorphisms) may occur in the equine karyotype without apparent phenotypic effect on health or fertility. Chromosome 13 is heteromorphic in normal horses for the size of the centromeric heterochromatin block (Buckland *et al.* 1976, Ryder *et al.* 1978). Nucleolar organizer regions (NORs) on chromosomes 1, 28 and 31 may be inapparent in some horses (Kopp *et al.* 1988). In 19 normal horses (Lippizaner, Hannoverian and Shetland) NORs were always present on both homologs of chromosome 1 and on at least one chromosome 31. Power (1988) reported that greater size variability exists in the horse Y chromosome than has been reported for the human chromosome. Comparing a group of 11 males with clinical abnormalities (e.g. XY sex reversal, infertility, congenital abnormalities, trisomy 28) to 20 clinically normal males, she found no apparent association of Y chromosome size with disease or infertility.

Sex chromosome abnormalities in females

63,X gonadal dysgenesis

The most common chromosomal abnormality (**X monosomy**) is associated with a missing sex chromosome and occurs among infertile mares (Bowling *et al.* 1987, Chandley *et al.* 1975, Hughes *et al.* 1975) (Figure 47). The karyo-

Figure 47: An Arabian mare with X monosomy was initially presented for karyotyping because of very small, inactive ovaries and primary infertility. Her height is under 14 hands (56 cm). Notice the acute hock angle ("sickle hocks").

type is similar to that described for human female gonadal dysgenesis, known as Turner's syndrome. Mares with chronic, primary infertility, failure to cycle regularly or at all and very small ovaries (0.5 × 0.5 × 1.0 cm) lacking follicular activity are candidates for chromosomal disease. Other characteristics of these mares are small size compared to expectation (say, 13:2–14 hands for an Arabian), over-angulated hind legs, bulky shoulders and large, wide-spaced ears. The condition occurs rarely and sporadically, probably from failure of the sex chromosome pair to separate (non-disjunction) during meiosis, producing one gamete without a sex chromosome (and another with two instead of one). In humans, X monosomy is a leading cause of early pregnancy loss and while the same may be true of horses, the condition has not yet been demonstrated in aborted horse fetuses.

No compelling evidence in humans or horses indicates an inherited tendency or association with maternal or paternal age at conception leading to offspring with X monosomy. Reports document its occurrence in dozens of breeds from ponies to draft horses. The mares obviously have no potential as broodstock, but have been reported to be fine riding horses and at least one was reported to be a champion cutting horse.

63,X/64,XX gonadal dysgenesis

In a few cases, karyotyping of mares with gonadal dysgenesis has revealed two cell lines, one with a normal set of female chromosomes, the other missing a sex chromosome. Mares with this mixed (**chimeric**) karyotype may rarely have foals, but infertility or, at best, subfertility is the most common finding.

64,XY sex reversal

The second most common chromosomal abnormality found among infertile mares is a male karyotype (Bowling *et al.* 1987). Race winners and show champion mares are among the cases identified in Thoroughbred, Quarter Horse, Appaloosa, Standardbred, Arabian, Morgan and Shetland Pony breeds. The reproductive phenotype is similar to 63,X gonadal dysgenesis, but XY mares do not show the poor conformation of X monosomy and may be taller than average for a mare of the given breed. This condition may be inherited, but probably more than one genetic type is involved, so the inheritance pattern will need to be determined for each family situation. Some types may be produced by X-linked genes (Kieffer *et al.* 1976) and others may be autosomal. In at least one group of Arabians the problem was clearly transmitted by a stallion to half of his sons (who appeared to be infertile females), but the genetic mechanism has not been determined (Bowling *et al.* 1987, Kent *et al.* 1986). At least one XY mare has had foals, but that situation should not be anticipated as generally likely (Sharp *et al.* 1980).

65,XXX

An extra X chromosome in female horses is associated with infertility and is phenotypically similar to XY sex reversal (Bowling *et al.* 1987, Chandley *et al.*

1975). The mares are tall, with scant or no palpable ovarian structures. From karyotyping case results reported worldwide, XXX is far less common in horses than X monosomy or XY sex reversal, yet in humans XXX is more common than either of these other conditions. In humans, XXX is not associated with infertility, another interesting point of comparative study for reproduction specialists since the only XXX cases so far identified in horses have been infertile.

Sex chromosome abnormalities in males

Karyotyping has not been as effective a tool for understanding infertility problems in males as it has been for females. Lack of examples of male chromosomal abnormalities may be correlated with the high incidence of early castration so that fertility potential is not evaluated in most males.

65,XXY

Only a single example of this karyotypic abnormality has been reported in horses, the counterpart of the relatively common Klinefelter's syndrome of male infertility in humans (Kubien *et al.* 1993). The horse with the abnormal karyotype was a grade draft horse, probably a cryptorchid, not a gelding, but case history details are vague since the horse was not examined by clinical specialists until about eight years of age.

XX male syndrome

A two-year-old bilaterally cryptorchid Quarter Horse colt, with a small but well developed penis and prepuce, mammary tissue at the sides of the prepuce and stallion-like behavior, was reported by Constant and colleagues (1994) to have the 64,XX karyotype of a female horse. Small testicles were located in the inguinal canals and testosterone levels (0.12 ng/ml) were slightly above the low baseline level of stallions (0.1 ng/ml). Based on cases of XX males in dogs and people, either a sporadic (non-inherited) or a hereditary etiology is possible. Sex reversal may occur from unusual recombination events between the pseudoautosomal (pairing) regions of the X and Y chromosomes (Kent-First *et al.* 1995).

XX/XY chimerism

McIlwraith and colleagues (1976) detected XX/XY whole body chimerism in a cryptorchid Arabian cross three-year-old colt with both testes undescended and not palpable in the inguinal canal, but fertility would not have been anticipated in such a case.

Ambiguous external genitalia (pseudohermaphrodite, intersex)

Ambiguity of phenotypic sex in the horse is associated with a variety of karyotypic abnormalities. The external structures are mostly described as female, with a penile-like structure in place of a clitoris, a (usually) blind vagina, gonads as intra-abdominal testicles, male behavior and XX sex chromosome constitution, such as the case described in an Arabian filly (Gerneke & Coubrough 1970). Similar cases have been diagnosed in other breeds, including Thoroughbred, Appaloosa and Paint. This phenotype resembles the inherited XX sex reversal syndrome seen in various breeds of dogs. Such cases in horses could also be associated with an inherited abnormality, but no data specific to that point have yet been presented for horses.

Other cases of intersexuality in the horse appear to be associated with more than one cell line and may best be explained as double fertilization or whole body chimerism arising from fusion of early embryos. A colt with a vulvar opening, right testis without germ cells, and XX and XY cells in karyotypes of lymphocytes and fibrous tissue was most likely a whole body chimera resulting from fusion of dizygotic twins in early embryogenesis (Basrur *et al.* 1970). Two intersex cases described by Dunn and colleagues (1981), one 64,XX/64,XY and one 63,X/64,XY, each had an underdeveloped penis, bilateral seminal vesicles, uterine tissue and bilateral ovotestes.

Chromosomal defects in early embryonic loss and abortion

In cattle and pigs lowered fertility can occur through early embryonic loss from chromosomally unbalanced gametes. Phenotypically normal parents heterozygous for chromosomal fusions between acrocentric chromosomes (**Robertsonian translocations** or **centric fusions**) have the potential for producing genetically unbalanced gametes during meiosis (chromosome non-disjunction). Translocation heterozygotes are reported among Caspian Ponies (Hatami-Monazah & Pandit 1979) but have not been directly implicated in infertility problems in that breed.

Power (1991) described an 11-year-old Thoroughbred mare with only two foals in eight years of breeding that had an abnormal karyotype consisting of heterozygosity for a balanced reciprocal translocation between two of the largest autosomes [64,XX,t(1q;3q)]. An eight-year-old fertile Thoroughbred stallion with a balanced tandem translocation between chromosomes 1 and 30 [63,XY, t(1:30)] was reported by Long (1994). Mares bred to this stallion would become pregnant, but lose the pregnancy in an early embryonic death. The translocation was transmitted to one of four offspring investigated, potentially continuing the embryonic loss problem in subsequent generations.

Blue (1981) attempted to establish tissue cultures of equine abortus material to obtain evidence for the frequencies and types of chromosomal abnormalities associated with equine abortion. He failed to establish a single culture, probably reflecting the generally poor condition of such material when received. Among 12 aborted fetuses successfully cultured by Haynes and Reisner (1982), no major structural or numerical abnormalities were found, but C-band polymorphisms were observed in two cases. The polymorphisms were also present in the normal parents so the significance of this finding is not clear.

Autosomal trisomies

In humans and cattle, extra autosomes are associated with such severe defects that the newborns seldom survive more than a few hours or days. An exception to neonatal death for autosomal trisomy is Down syndrome in humans, a trisomy of chromosome 21. Horses also show (rare) viable trisomies. An extra autosomal chromosome has been identified in four yearlings, not recognized for infertility, but described as small, with stiff gait and poor conformation (Bowling *et al.* 1987, Bowling & Millon 1990, Klunder *et al.* 1989, Power 1987). Two were Thoroughbreds, one an Arabian and one a Standardbred. Two were females and two were males. In each case a different small chromosome was involved (chromosomes 23, 26, 28 and 30) (Figures 48 and 49). At least two were offspring of older mares, perhaps paralleling the association of Down syndrome with increased maternal age at conception.

Figure 48: A Thoroughbred mare with trisomy for chromosome 26. She was presented for chromosomal evaluation because of poor conformation, unthriftiness and lethargy.

Figure 49: An Arabian filly with trisomy for chromosome 30. She was presented for chromosomal evaluation because of small size and poor conformation. She was an otherwise active filly with a bright attitude.

Karyotyping services

Chromosome analysis requires a tissue sample with actively dividing cells. This criterion eliminates autolyzed tissues, conventionally frozen tissues, frozen semen samples prepared for future use in artificial insemination programs, fixed tissues and probably even refrigerated tissues as a source of material for karyotyping. Fresh blood samples are the most often used tissue source. A blood sample can be sent by overnight courier to a laboratory specializing in animal karyotyping. The laboratory will isolate lymphocytes, then incubate them in a tissue culture medium containing substances that stimulate the cells to divide. After three days, the lymphocytes are harvested, fixed, put on slides and stained with dyes that bind to chromosomes. Using a micro scope, the slides are scanned for cells in which the chromosomes are contracted and well spread apart. Suitable cells are photographed. The individual chromosomes from a single cell are cut out from the photographic print and aligned in a paired array known as a karyotype. (Computer-assisted video imaging technology is now often being applied to avoid the time-consuming steps associated with conventional photography.) It is from the paired array that the diagnosis of missing or extra chromosomes can be obtained.

CHAPTER 13
Genetics of performance traits

Many horse breeds are distinguished by a particular performance ability achieved in concert with a human handler. These breed hallmarks were established prior to an understanding of genetics or modern breeding theories. Thoroughbreds have been intensively selected for more than 300 years for racing speed at the gallop. Other breeds have been selected for racing speed at the trot (e.g. American Standardbred, Russian Orlov, Finnish, North Swedish). Stock breeds (e.g. Quarter Horse, Paint, Appaloosa) have been bred for ranch and cattle work and also excel at short distance racing. Draft breeds (e.g. Shire, Clydesdale, Percheron, Belgian) have been selected for pulling ability. Icelandics, Paso Finos, Peruvian Pasos, and Tennessee Walking Horses are bred to perform riding gaits, probably of pre-historical origin but considered specialized compared to the walk, trot, and canter repertories of most horses today. In Europe and Asia horses may be bred for meat or milk, but those traits are not major goals for most horse breeders.

A unifying theme in horse breeding is that the purposes for which most modern horses are bred require moderate to extreme athletic ability and the savvy to interpret and obey instructions from their human companions. The close relationship of the horse to a human handler distinguishes breeding goals for horses apart from other livestock species. That special abilities are inherited is intuitively clear, but none seems to be inherited as a simple genetic trait. Performance characteristics may be controlled by genes at several loci acting together in an additive fashion (quantitative trait loci). Quantitative traits are measurable and influenced by the environment (e.g. nutrition, trainer, rider, weather) as well as by genes. Our comprehension of the inheritance of performance traits strongly depends upon our ability to identify and measure genetic and environmental effects as separate components.

A discussion of horse performance genetics must be largely theoretical due to our present inability to precisely identify the various trait components and to predict the outcome of their complex interactions. The information is slanted toward racing performance, since a large number of racing records are available to derive and test quantitative trait theories. For further information about the general field of animal breeding theory, Pirchner (1983) is an authoritative and comprehensive text book. Van Vleck (1990b) provides an excellent discussion of selection theory with examples specifically related to horses. Comprehensive reference lists to consult for information on horse performance genetics include Hintz (1980), Tolley and colleagues (1985) and Klemetsdal (1990).

Performance breeding plans

The successful breeder is one who can identify and combine the available components of genetic and environmental excellence. Schemes to select breeding stock for a performance trait assume that the stock is genetically variable. If all the variation in performance were due to environmental factors, then genetic selection of breeding stock would be ineffective. Selection schemes also assume that the best performers have the desirable genes. Since environmental factors can enhance or mask genetic effects, the breeder needs to know to what degree performance excellence is inherited. For simple modeling, breeding schemes assume that the trait shows additive genetic variance, although other possibilities such as overdominance (heterozygous advantage) could also be involved.

To benefit optimally from genetic selection theory the breeder must set environmental components to be maximally advantageous and:

- have breeding stock with genetic variation for the trait
- know the accurate predictors of performance excellence
- minimize generation intervals
- practice high selection intensity.

Genetic variation in performance traits

The measure of the genetic component of phenotypic variation is called **heritability** (h^2). A highly heritable trait will have a value near 1.0. Most published reports on performance traits in horses have estimated moderate to low heritability values, ranging from about 0.5 to near zero. Various methods are used to estimate heritability such as offspring–sire or offspring–dam correlations or paternal half-sib correlations, and generally include adjustments for sex and for environmental (non-genetic) biases such as age.

A trait may be highly heritable in one population and only moderately heritable in another due to differences in genetic and environmental background. Efficient and long-term selection practices in one breeding population may theoretically lead to genetic uniformity, so that the variation observed in a measured trait may be due essentially to environmental effects. Association of genetic components of performance excellence with blood typing markers, electrocardiogram measurements, physiological, skeletal or muscular characteristics is largely unstudied. Heritability has been estimated both for objectively evaluated traits,.such as speed, jumping height and pulling ability, as well as for subjectively evaluated ridden show ring and working events.

Working performance (pulling ability and cutting)

As reviewed by Hintz (1980) selection for pulling ability in Finnish draft horses, based on information on progeny or collateral relatives and the animal's own record, is expected to be moderately effective since heritabilities for amount of weight pulled, endurance in pulling, way of straining, and way of pulling have values in six studies from 0.23 to 0.27.

Cutting ability or "cow sense" in horses has historically been an important trait sought by stockmen who manage range cattle on horseback in western North America (Figure 50). Based on scores obtained during National Cutting Horse Association Futurity competition, Quarter Horse sires were evaluated through the performance of their offspring (Ellersieck *et al.* 1985). Heritability was estimated to be 0.19 ± 0.05. However, since the horses competing were not a random set of cutting horses, but rather the selected elite, this estimate may be biased downwards.

Figure 50: Cutting ability in Quarter Horses had a low heritability value in one study, but the horses used for the estimate were an elite subset of cutting horses—performers—so this value may not reflect the heritability in the breed at large.

Riding performance (jumping, three-day event, and dressage)

Analysis of heritability of riding performance has been most frequently studied in European warmblood breeds, for which results are readily available from performance testing of three-year-old stallions. Moderate to high heritability values were obtained for traits such as riding ability (0.36), gaits (0.50), jumping ability (0.72), cross country (0.33), racing time (0.53) and character/temperament (0.25) using the results of German stallion performance tests (Bruns *et al.* 1985, cited in Klemetsdal 1990). Estimates of genetic merit for riding performance can also be based on such data as competition scores awarded by judges, earnings, or placings. Age, sex, and height of the animal may influence riding performance and need to be considered when obtaining heritability estimates from competition data. Hintz (1980) averaged heritability estimates based on paternal half-sib comparisons of the log of earnings from several studies for jumping (0.18), three-day event (0.19) and dressage (0.17) in French, German and Swiss Warmbloods. These data suggest a low heritability for the complex trait of riding ability in these breeds.

Racing performance

A great abundance of data is available for racing performance at the trot, pace, or gallop and the heritability of racing ability has received more attention than any other performance trait of the horse. Among the possible measurements of racing success are earnings, log of earnings, average earnings index, lengths behind winner, performance rates, handicap weight, time of finish, average time, time as a deviation from winning time or average time, best time, rank at finish, and percentage of races won. Hintz (1980) averaged heritability estimates from a number of studies of Thoroughbreds for log of earnings (0.49), earnings (0.09), handicap weight (0.49), best handicap weight (0.33), time (0.15) and best time (0.23); of trotters for log of earnings (0.41), earnings (0.20), time (0.32) and best time (0.25); and of pacers for best time (0.23). Among the environmental (non-genetic) effects which introduce bias into estimates of the genetic component of racing excellence are age, sex, non-random mating practices, class of race, track condition, distance, and trainer and jockey effects.

Conformation

Measurable traits such as height, girth and cannon bone circumference are generally considered to be moderately to highly heritable. Literature searches provide a few citations (e.g. Legault 1976), but breeds are not specified. Heritability of conformation traits is clearly an area open for further research, particularly for a variety of breeds and types.

Predictors of performance excellence

A simple scheme for breeding animals of merit relies on identification of superior breeding stock using the measurement of traits with a moderate to high heritability. Most quantitative traits have heritabilities between 0.1 and 0.5. In other terms, these heritabilities translate to a prediction of genetic merit for which accuracy ranges from 32 to 71%. Using a trait for genetic selection, such as log of earnings in Thoroughbreds, whose heritability is moderately high at 0.41, selection could theoretically be effectively made on the record of the performer without the need for collateral information from relatives. For traits with lower heritabilities, progeny records and those of close relatives can increase accuracy of prediction. Stallions probably have greater accuracy of prediction than mares because they have their own records, as well as the records of more progeny.

More accurate predictive models of genetic excellence could establish indices of breeding value based on incorporating parameters to account for various sources of environmental bias. Genetic improvement to increase production traits (milk, meat, eggs and fiber) in other livestock species has most recently been making progress with the application of a quantitative genetics model known as BLUP (best linear unbiased prediction) developed by C.R. Henderson. Klemetsdal (1990) considers BLUP to be the method of choice in predicting breeding values in horses because the animal model form of BLUP should maximize the accuracy of breeding values and the expected genetic gain from selection, as well as yield unbiased estimators of genetic and environmental trends in the population. The usefulness of BLUP for predicting breeding values of horses should become obvious in the next decade because it will surely be tested on a variety of traits in a number of breeds and populations.

Generation interval

The generation interval is the average age of the parents when their offspring are born. The generation interval influences the expected rate of genetic improvement per year for quantitative traits. Horses have a relatively long generation interval of 9–11 years compared with other domestic animals, whose values are half that or less (Pirchner 1983). Often generation intervals for sires and dams are different, complicating the estimation of genetic progress. One disadvantage to lowering the generation interval is that the accuracy of prediction might decrease, since fewer records are available for use in estimating breeding value.

Selection intensity

If only animals with the highest genetic value are used for breeding, then the average merit of the population must improve. The rate of improvement depends upon the intensity of selection. Usually culling is most strongly practiced for stallions in which only the top few percent of the performers are used for breeding, and selection is not usually as intense for mares. For example, Gaffney and Cunningham (1988) estimated selection intensities in Thoroughbreds as 6% in males and 52% in females. In other words, 94% of Thoroughbred colts and 48% of fillies do not contribute genes to the next generation.

"Cunningham's paradox"

Cunningham (1975) showed that winning times in classic Thoroughbred races in England have made only a slight improvement during recent decades, despite the fact that heritability estimates of winning time are generally in the moderate range. Suggestions to explain the paradox include the possibility that Thoroughbreds have reached their highest level of genetic potential or that estimates of heritability values are too high due to the inability of estimates to include assortative mating practices, the high homogeneity of mare bands, and the distribution of environmental effects according to the quality of the parents (Langlois 1980).

An analysis of the Thoroughbred population as a whole, not just winners in the best races, showed genetic gain for stallions born from 1952 to 1977 using BLUP analysis for TIMEFORM handicap ratings (Gaffney & Cunningham 1988). The amount of progress was estimated to be 0.94 ± 0.13 TIMEFORM units per year, which correlated with the predicted rate of genetic change based on an estimated heritability value of 0.36, selection intensity of 6% for males and 52% for females, and generation intervals of 11.2 ± 4.5 for sires and 9.7 ± 3.1 for dams. These data suggest that the Thoroughbred population still contains genetic diversity with respect to racing ability and the reason for lack of progress in winning times of the principal classic races will need to be sought elsewhere.

CHAPTER 14
Pedigrees and breeding schemes

Breeders cannot change Mendelian genetics, nor the number of genes involved in traits, nor their linkage relationships. They cannot change the physiological interactions of gene products, but they can hope through selective mating to produce gene combinations that consistently result in high quality stock.

The formation of breed societies to record animal pedigrees, the foundation of Western livestock breeding practices, traces to England in the early nineteenth century. The breed societies aimed to protect and promote a distinctive animal type consistently superior in production or performance characteristics compared to common stock. The achievements of these societies lent credence to the notion that "pure" stock, whose genealogy was faithfully recorded and published, was highly desirable for successful animal breeding. The availability of recorded pedigrees, presumably authentic, led to attempts to use them to correlate success with pedigree patterns and as a tool to predict the outcome of matings.

Broadly speaking the systems of mating that a breeder may choose are:

- Mating like to like (based on pedigree likeness or individual likeness, such as performance success, disposition or body shape).
- Random mating (no selection).
- Mating unlike to unlike (based on outcrossed pedigrees or individual extremes, such as tall with short, or rangy with compact, or hot temperament with mellow).

Successful examples of all these schemes could be cited for any breed. Every horse breeder needs to understand the genetic principles underlying these situations and then decide which is the most appropriate for each breeding pair.

This chapter discusses terms and concepts of pedigree study. One aim of teaching is to encourage debate and critical examination of issues and statements. This chapter challenges some of the myths of horse breeding. If it provokes discussion and sound research to help breeders make wise choices, it will have served one of its intended purposes. For additional discussion, consult Lush (1945), a classic, clearly written animal breeding text.

The pedigree format

We use pedigrees to help us understand what genes a horse may have. Mostly what we can learn from a pedigree provides us with a subjective impression, although some "real" genetic information can be learned. For example, a tobiano horse has no possibility of being homozygous for tobiano if it does not have both a tobiano sire and a tobiano dam. Notice however that if the tobiano does have two tobiano parents, the pedigree cannot tell us whether the offspring has one or two tobiano alleles.

Standard diagram

The routine pedigree format is a diagram of a "begat" listing. By convention, the offspring is named to the left of the page and the first column to the right lists the parents, arranged with the sire on top (Figure 51). The next column gives the parents of the parents (the grandparents of the offspring) and so on. The pedigree may be a listing of names, or names and registration numbers, or may include as well year of birth, coat color, breeder, photographs and outstanding performance records. The more information provided, the more accurate will be an estimation of the individual's characteristics.

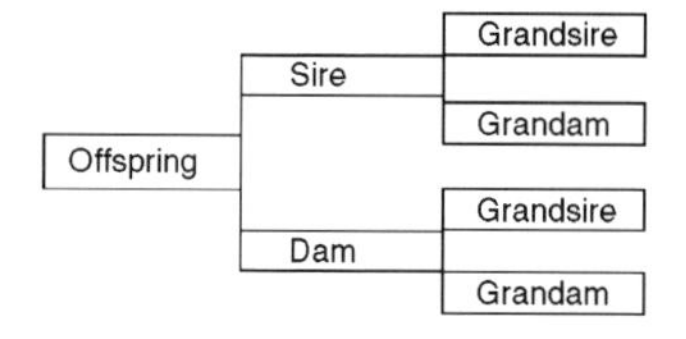

Figure 51: Conventional pedigree format, providing a short listing of the parents and grandparents of a particular offspring. A typical pedigree will provide five or six generations, rather than the two given here. The more information provided about each of the ancestors, the more useful will be the pedigree for predicting the potential of the offspring.

Relatedness and "percentage of blood"

Offspring resemble their parents to varying degrees, but the proportional genetic contribution of each parent is constant: half the genes of an offspring come from the mother and half from the father (except for mitochondrial genes that are maternally inherited in a non-Mendelian fashion). **A relatedness coefficient** of 50% or 0.5 is assigned to the parent–offspring relationship.

Full siblings *on average* share 50% of their genes. For every locus, assuming the parents are heterozygous for different alleles, the offspring have four possible allelic combinations: 25% of the time they will have received the same alleles from their dam; 25% of the time they will have received the same alleles from their sire; 25% of the time they will have received no alleles in common from sire or dam; and 25% of the time they will have received the same alleles from both sire and dam. The random assortment of chromosome pairs during gamete formation means that we cannot predict the exact proportion of genes that any two full siblings have in common; we can only provide an average value for all genes of full siblings as a group. In practical terms, stallion advertisements to the contrary, one cannot assume that a full brother to a proven stallion will be an equivalently successful sire. Certainly, he has a greater likelihood of sharing genes with the proven stallion than does a non-relative, but there is no guarantee that he has received the group of genes that sets his brother apart from the rest of the breed.

Half-siblings on average share 25% of their genes, and first cousins share about 12.5%. Relatedness decreases by half with each succeeding generation. More complicated relationships can easily be calculated.

In the simplest of pedigree evaluations, breeders may talk about "percentage of blood." Of course blood is not a unit of inheritance, but is used in this context to imply genetic traits. Summing the relatedness coefficients for every occurrence of particular ancestors in the pedigree provides the proportion by source for an individual's genes (Figure 52). The sum is not an exact proportion, but a statement of the most likely percentage. It is always possible that the gene proportion could be larger or smaller than the calculated relatedness coefficient.

Evaluation of pedigree influences

A four- to five-generation pedigree with its 30–62 names among the various generations can be intimidating to breed newcomers to whom few of the names have any recognition value. It is natural to seek simple rules to reduce the blur of names to a meaningful collection. Some breeders believe that horses beyond the fourth generation in a pedigree need not be considered. This statement is given to justify the unimportance of pedigree elements generally considered unfavorably. Other breeders use pedigrees as "witch hunts" and will not use a horse whose pedigree has *any* perceived undesirable element. Usually, this thinking is associated with trying to avoid genetic diseases associated with named horses. The balanced view of pedigree evaluation does not use simple generalizations, but requires thoughtful analysis on a case-by-case basis.

The genome of any horse is always a composite of contributions from the entire pedigree. The fraction of genes in an offspring's genome attributed to particular ancestors in distant generations is small. Any single gene in a fifth generation ancestor has only about a 3% chance of passing through all the meiotic partitions and arriving in the genome of the great-great-great grandchild. Yet consider a pedigree from the standpoint of a single color gene, such as gray, which can be easily traced through a pedigree. STAR in the illustrated pedigree is a gray, with two potential pedigree sources of the dominant allele *G*. The gray horse on the sire's side cannot be the source of STAR's gray color, since the color is not even transmitted from that gray ancestor to the next generation. The gray ancestor on the dam's side is clearly the source of the color for STAR. With every generation the *G* on the dam's side had a 50%

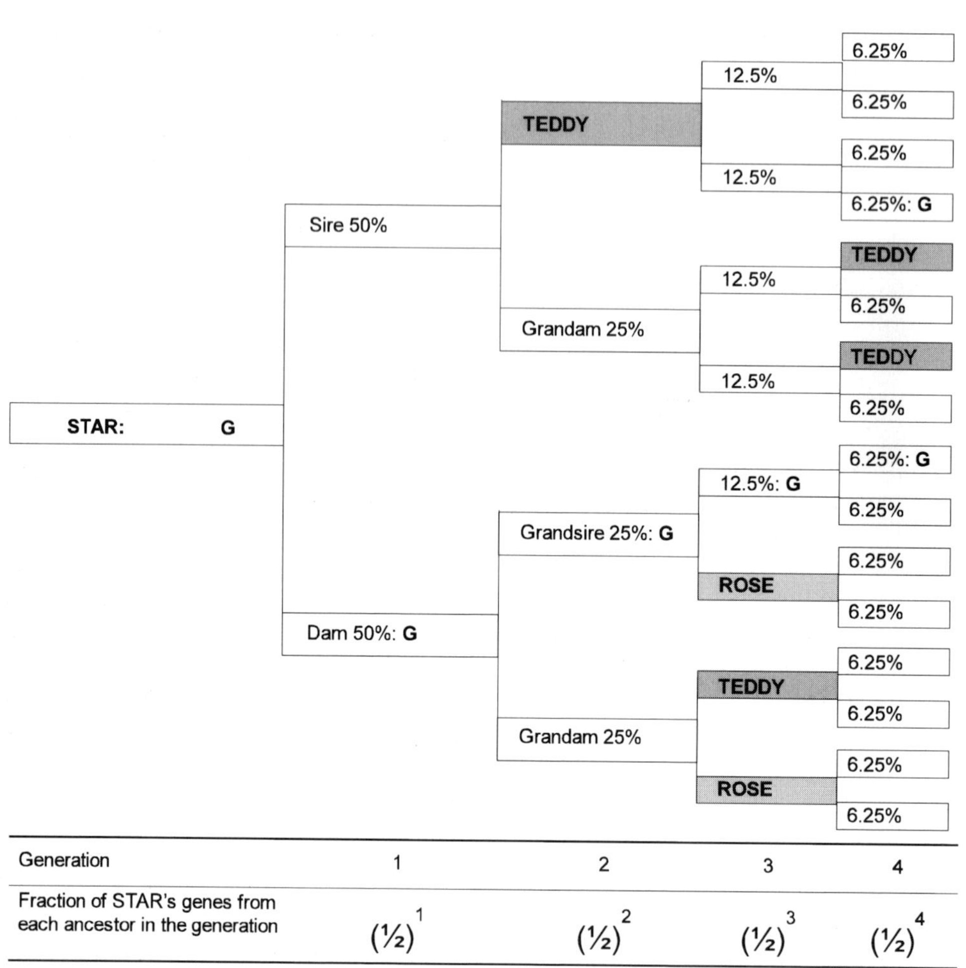

Figure 52: The fraction of an individual's genes most likely to come from each ancestor is presented in this conventional pedigree format. For a hypothetical case, consider the calculation for a pedigree of STAR. Summing the fractions, this pedigree could be described as "50% TEDDY," meaning that *on average* an individual with this pedigree could be expected to have 50% of its genes derived from TEDDY, but it could also have a smaller or larger proportion. STAR has a relatedness coefficient of 25% to the mare ROSE. Horses of gray color are designated with G.

chance of being lost from this pedigree, but in each case it was transmitted forward. The gray color example shows it is not the position in the pedigree that determines whether traits are transmitted, it is the chance events of probability.

Relate this color example to genes that are not shown on a pedigree, say disease genes. Among ancestors in the same generation with the same heterozygous pair of alleles for a given gene, it may be very difficult to determine which of their allelic alternatives was transmitted when their presence is not phenotypically obvious. A single disease or color gene could be traceable to a sole source in each generation. On the other hand, multiple elements may be possible sources of the gene. If a trait is known *not* to have been transmitted to the next generation, it is a waste of time to continue to consider that individual or pedigree line as a source of the given desirable (or undesirable) gene.

Inbreeding and linebreeding

Inbreeding pairs related animals, such as father and daughter, mother and son, brother and sister, or cousins. STAR (Figure 52) and his sire are inbred to TEDDY. STAR's dam is inbred to the mare ROSE, but STAR is not inbred to her, because she does not occur on both sides of his pedigree.

Linebreeding is the term used by breeders to describe their programs based on multiple pedigree crosses to a single exceptional animal. STAR is linebred to the "highly renowned" TEDDY.

An **inbreeding coefficient, *F***, provides a probability that the alleles in any gene pair will be *identical by descent* (homozygous) from an ancestor found on both the sire's and dam's sides of the offspring's pedigree. Examples of *F* values for matings in human pedigree terms are: parent–offspring or full-sibs (0.25); uncle–niece (0.125); first cousins (0.0625); second cousins (0.016). STAR's *F* value is 0.09375, meaning about 9% of his genes would be homozygous for alleles originating from TEDDY. Several inbred relationships may occur within a horse pedigree. In those cases an inbreeding coefficient is calculated by summing the coefficients of all the relationships.

In general horse breeding programs, calculation of inbreeding coefficients probably contributes little of practical value. For intensely inbred breeding groups in which the immediate aim is to select matings to maintain a maximum of genetic diversity (say, for an endangered species), selection for lower inbreeding coefficients could be useful to choose among breeding pair options. Readers interested in learning to calculate such values using examples of horses may want to consult Van Vleck (1990a).

Toward homozygosity

The theoretical purpose of inbreeding (or linebreeding) is to produce stock of consistent excellence through enhanced relationship to admired individual ancestors. Inbreeding not only increases the proportion of genes that trace to a given ancestor, it also increases the likelihood that the genes will be homozygous. The premise underlying the positive attitude toward inbreeding is that homozygosity is desirable, but these genes may *not* have been homozygous in the admired ancestor. Key components for working with inbred pedigrees to avoid producing foals homozygous for an undesirable trait include accurate tests to identify trait carriers and knowledge of basic genetics. For example, inbreeding to a palomino Quarter Horse stallion, highly esteemed for his conformation as well as color, could lead to the production of cremello foals that would not be eligible for registration regardless of desirable conformation and quality. Breeders could avoid the cremello problem and still inbreed to a palomino source if they apply principles of coat color genetics and select appropriate breeding pairs (avoid breeding together palominos, buckskins or their combination). This example illustrates the possibilities of using basic genetics to avoid producing foals homozygous for an undesirable trait while working with heterozygous (carrier) pedigrees.

Although physical uniformity may be a goal of purebred breeding programs, extreme genetic uniformity is probably not desirable, because hardiness, vigor and reproductive soundness may decrease with increasing homozygosity. In a study of fertility and inbreeding in Standardbreds, reproductive performance was not affected at the low inbreeding levels of 7–10% typically encountered in domestic horse breeding programs (Cothran *et al.* 1984). The effects of high levels of inbreeding on domestic horse fertility have not been reported.

Genetic alternatives will become restricted as homozygosity increases in any inbred group. Inbreeding tends to create distinctive breeding groups through chance "fixing" of genes. If several breeders undertake inbreeding programs independent of each others' stock, different genes can become fixed in each group, although their genetic repertories might initially have been quite similar. *Diversity can be maintained within a larger (breed) context by establishing several inbred lines.*

Closed studbooks

Many studbooks have regulations that prevent the use of animals outside the registry, effectively creating a closed gene pool. The aim of this closure is to encourage breeding of a consistent type of stock with excellence for a selected set of breed characteristics. For closed studbooks, the gene pool is initially defined by variants provided by the founder stock. With a sufficiently large founder population and no historical periods of severely restricted population size (**bottlenecks**), the gene pool may be quite diverse. New gene combinations may still be realized for tens of generations.

The only sources of new genetic material will be mutation or undetected crossbreeding (**gene introgression**). The mutation process occurs spontaneously and continuously, but most mutations are immediately eliminated because they are deleterious to the organism or eliminated within a few generations by random chance. Introduction of new genetic material into a closed studbook through crossbreeding can be avoided by denying registration to horses that fail parentage verification tests.

Maintenance of diversity may be an appropriate breeding goal, particularly within the context of a closed breeding program. Several horse breeds are fractionated into subsets of inbred lines, sometimes defended so vehemently by their protagonists that personal feuding between owners of different substrains becomes surprisingly intense and bitter. The far-sighted view of breed promotion recognizes the value of maintaining multiple lines to preserve diversity for the genetic health of the breed.

Inbreeding in short *vs.* long pedigrees

Owners generally see no more than a five- or six-generation pedigree of their animals, owing perhaps to constraints as trivial as paper size for printing pedigrees, but also to the time necessary to research and write out more generations. If distant generations could be viewed as easily on paper as the more immediate ones, breeders might be surprised to find out to how few founders their purebred animals trace. In a study of inbreeding and pedigree structure in Standardbred horses, MacCluer and colleagues (1983) determined that inbreeding coefficients calculated from 27-generation pedigrees (back to breed founders) were increased by nearly 2.4 times (from 0.0375 to 0.0888) compared with those calculated from the conventional short pedigrees usually considered by breeders.

Outcrossing

In contrast to inbreeding or linebreeding schemes, some breeders use outcrossing programs to meet their goals. Outcrossing means the breeding together of unrelated animals. Outcrossing in the context of a closed breed is practiced by avoiding or minimizing the duplication of names in pedigrees, although it is likely that the animals are related in distant generations.

Some breeds allow the use of outcrossing with horses not in their studbook, usually only to selected breeds considered important sources to obtain or maintain performance or conformation characteristics. In theory, horses that are the products of outcrossing programs (also called crossbreeding) may be more heterozygous than inbred horses. Crossbreds might be outstanding performers, but as breeding stock inconsistently pass on their desirable characteristics. Warmblood breeds extensively use outcrossing in combination with performance testing schemes to maintain a breeding stock pool strongly selected for a subset of traits associated with predictable excellence in three-day events and selected specialties.

Assortative mating

Mating like to like on the basis of their perceived similarities, without regard to pedigree, is called **positive assortative mating**. This technique could combine animals with similar genes or animals with different genes effecting a similar phenotype. In any case, the goal is to produce an animal that closely resembles the parents.

Mating together animals of unlike phenotypes is called **negative assortative mating**. The goal of this process is to breed offspring that are not as extreme as either parent.

The consideration of pedigrees from the standpoint of whether they represent positive or negative assortative matings can be useful for thinking about individual traits, but probably has limited application in the context of the overall horse. A racehorse breeder would probably select positive assortative mating for speed (sprinter to sprinter or distance winner to distance winner), but negative assortative mating if seeking to overcome conformation defects (a heavier boned mate to improve light bone). Most matings combine positive assortative mating for some traits with negative assortative mating for others, but in practice breeders seldom articulate their breeding decisions in such terms.

Tail-male lineage

Breeders may classify breeding lines by tail-male pedigree lineage tracing the top line of a conventionally drawn pedigree back to a noted founder stallion (Figure 53). Clustering male horses based on their shared tail-male relationship makes biological sense in the light of transmission of the Y chromosome from male to male. These horses necessarily share the few genes present on the Y chromosome (barring the effects of recombination and mutation).

Figure 53: SAM, TONY, DAVE and MIKE share tail-male lineage and a Y chromosome from the stallion BUDDY.

From a genetic standpoint it is not clear how females could be considered to have a connection to a tail-male group since they do not have the founder Y chromosome. After several generations both males and females may have no autosomal genes remaining from that founder source.

Tail-female lineage, strain breeding and families

Several meanings are associated with the terms strain and family when applied to pedigree relationships. Probably the most common use for either term (and they may be used interchangeably) is for a group of animals selectively bred for several generations from a subset of animals within a breed. Often this usage reflects the long-standing successful breeding program of a particular stable, stud farm or group of cooperative breeders.

A different meaning is attached to these terms by some Thoroughbred and Arabian horse breeders, who tie physical and mental traits to tail-female pedigree connections (Figure 54). Bruce Lowe, an Australian who made

Figure 54: SAM, JANE, DAVE and SPOT share tail-female lineage to the mare LADY.

extensive studies of Thoroughbred pedigrees, developed a system for predicting racing excellence based on tail-female founders (Bruce Lowe families). The Bedouin Arab used the tail-female convention to describe the relationship of their horses to celebrated foundation mares (strains), a tradition still continued by some breeders.

The performance excellence of horses sharing tail-female connections could be based on maternal inheritance through mitochondria, but any such relationship between performance traits and specific mitochondrial genes currently awaits validation. Some breeders strive for multiple generations of breeding within the same strain (sires and dams of the same strain), but the underlying biological basis for attaining excellence in this way is not clear.

A tail-female connection among female horses is not chromosomally analogous to the Y-chromosome sharing among males with a tail-male relationship. Any grandchild of our hypothetical LADY has only a 50% chance of receiving one of LADY's X chromosomes. Two otherwise unrelated granddaughters (or grandsons) have only a 12.5% chance of receiving the same X chromosome from their common maternal grandmother. Thus the persistence of traits through pedigrees cannot be routinely traced to X-chromosome sharing from a tail-female connection.

Questioning breeding myths in the light of genetics

Newcomers to horse breeding often look for pedigree formulas or hope to emulate a particular breeder's program by using related stock. Unfortunately for novices, the truths of horse breeding are that many successful horse breeding judgments are in equal measure luck and intuition. Horse breeding is not as easy to fit to formulas as breeding for meat or milk production. Many of the highly valued traits of horses, such as breed type or way-of-going, are subjectively evaluated in show ring events. Winners may reflect the skills and show ring savvy of the trainer/handler, as much as the innate abilities of the horse. Some breeders can learn to predict to their satisfaction the approximate phenotype to expect from a selected mating because of their years of experience studying horses and their pedigrees, but their skill cannot always be taught to others and may not work with unfamiliar pedigrees.

Nicks

Horses considered to be of excellent quality often present a pattern of recurring pedigree elements. Breeders naturally seek to define pedigree formulas or "nicks" to design matings that will consistently replicate this quality. But breeding horses is not like following a recipe to make a cake. You cannot precisely measure or direct the ingredients (genes) of the pedigree mixture as you can the flour, sugar, chocolate, eggs and baking powder for a cake. You can construct pedigrees to look very similar *on paper*, but the individuals described by those pedigrees may be phenotypically (and genetically) quite different. Before seriously considering any breeding formula scheme it is essential that breeders understand the most basic lesson of genetics: *each mating will produce a genetically different individual with a new combination of genes.*

A certain nick is often expressed as cross of stallion A with stallion B—an obvious impossibility! Probably one source of this convention is that it is easier to become familiar with the characteristics of the offspring of stallions than mares, because they usually have a great number of foals. Another source is the perceived need to reduce complex pedigrees to an easily described summary. Breeding stallion A to *daughters* of stallion B (this would be the genetically correct description of some nick) may produce horses of a relatively consistent type compared with the rest of the breed. For mares of the next generation, the "magic" nick (stallion C) is again at the mercy of genetic mechanisms that assure genes are constantly reassorted with every individual and every generation. Some breeders are reluctant to introduce stallion C at all, preferring to continue with their A–B horses, breeding their A–B mares to A–B stallions. If a nick works, and it can appear to do so for some breeders, basic understanding of genetics tells us that it is seldom a long term, multi-generational proposition unless it is guided by an astute breeder who is making breeding decisions on individual characteristics, not merely the paper pedigrees.

Basing a program on champions

Novice breeders are often counseled to "start with a good mare." This seems to be reasonable advice, but does not make it clear that the critical point is learning to recognize a good mare. Sometimes breeders fail to produce a foal that matches the quality of its excellent dam, while less impressive mares in other programs produce successfully. Probably the lack of objective criteria to evaluate horses accounts for both observations. A "good mare" need not be a champion, and a champion is not guaranteed by dint of show ribbons to be a "good mare." In addition, we do not know the inheritance patterns of highly valued traits for show ring excellence. If the ideal type is generated by heterozygosity (e.g. the ever useful example of palomino), the only infallible way to produce foals that meet the criterion of excellence (palomino color) is to use parents of less desirable type (chestnuts bred to cremellos). This example is not to be taken as a general license to use horses of inferior quality, but to provoke critical thinking about the adequacy of general breeding formulas to guide specific programs.

Other breeders pride themselves in structuring programs based on using exceptional stallions. However, breeders should be aware of the fallacy of this type of strategy: "I like stallion Y but I can't afford the risk to breed my mare to an unknown stallion like Y—I can only breed to a National Champion like Z." *Any* breeding is at risk to produce a less than perfect foal, but the advertising hyperbole leads novices to think that certain avenues are practically foolproof. Included in the best thought out breeding plans must be an appreciation of the ever-present potential of deleterious genes being included with those that are highly prized. It is irresponsible to assume that an animal is without undesirable genes. The wise breeder understands the task as minimizing the risk of creating a foal with serious defects and maximizing the chances of producing an example of excellence.

A master breeder needs several generations (a generation interval for horses is estimated to be 9–11 years) to create a pool of stock that contains the genetic elements that he or she considers important for the program vision. To learn to identify essential characteristics, a breeder needs to evaluate the horses and their pedigrees, not advertisements or pictures. When a breeder discovers those elements, he or she can make empirical judgments and is on an obvious path for making good breeding decisions.

The cult of the dominant sire

In some circles, the highest praise of a breeding stallion is that he is a dominant sire. Another widely encountered livestock breeding term for an elite sire is prepotency. The implication is that all his foals are stamped with his likeness, regardless of what mare is used. This concept would appear to contradict the advice to "start with a good mare." Those owners who strongly believe in the strengths and qualities of their breeding females would surely question the value of a so-called dominant sire who could seemingly obliterate valued characteristics that would be contributed by their mares.

A good understanding of genetics should allow a breeder to put the proper frame of reference to terms such as dominance and prepotency as applied to breeding horses. Some animals transmit certain characteristics at a higher frequency than is generally encountered with other breeding animals. Coat color is always the conspicuous example. Any stallion whose offspring always or nearly always match his color is popularly described as a dominant sire. To be exactly correct, for at least some of the effects being considered, the genetic interaction is not dominance but epistasis and homozygosity. A stallion could be homozygous for gray, leopard spotting or tobiano, so that every foal, regardless of the color of the mare (with the possible exception of white), would have those traits. Homozygosity for color is not necessarily linked with transmission of genes for good hoof structure, bone alignment in front legs, shoulder angulation or other traits that may be desirable. Most conformation traits seem to be influenced by more than one gene. Some stallions may be exceptionally consistent sires of good conformational qualities, but it is unlikely that every foal will have these traits or that any stallion could be so characterized for more than a few traits.

The balanced view is that a battery of stallions is needed to meet the particular genetic requirements of each of the various mares in the breed. No one stallion can be the perfect sire for every mare's foal.

Using genetics to guide a breeding program

If assays for genes important for program goals are available, the probability of obtaining foals with selected traits from specific breeding pairs can be predicted. For many horse coat colors, offspring colors can be predicted, but conformation and performance traits are not well enough defined for predictive values to be assigned. So little is known about the genetics of desirable traits, it is premature to suggest that any general technique of structuring pedigrees consistently produces either better or worse stock.

The important lessons to learn from genetics to use for horse breeding decisions may seem nebulous to those looking for easy "how-to" information. Yet an appreciation of how genes are inherited, the number of genes involved in the makeup of a horse, their variability within a breed and the inevitability of genetic trait reassortment with every individual in every generation will provide the critical foundation for sound breeding decisions.

With the current interest in genetics and the new technologies available for looking at genes at the molecular level, information about inherited traits of horses is likely to increase significantly in the next decade. Horse owners can help with the process in several ways, including communication with granting agencies about specific problems of interest to them, providing money to fund the research, and providing information and tissue samples to funded research studies. Horse breeders are eager to have sound genetic information and diagnostic tests to guide their programs and, fortunately, the future looks very promising.

CHAPTER 15
Sorting out factors in development: genetics vs. environment and sire vs. dam

Genes provide the blueprints for development and growth of a foal, but non-genetic environmental factors occasionally modify actions of genes. Breeding good horses requires making selection decisions based on trait genetics and providing the optimal environment for gene action. Sorting out the relative roles of genes and environment for selected traits can be a complicated proposition.

When family members share an environment, the effects of non-genetic factors may mimic the appearance of an inherited trait. Analysis of a trait is also complicated by the fact that a rare few inherited genes do not show Mendelian inheritance patterns. Horse breeders will want to understand these situations, even if they happen so infrequently that few examples from horses can be given. Related to this discussion is the recurring theme among horse breeders that the dam is more significant to the overall worthiness of the foal than is the sire. This chapter provides a brief look at the issue of genetics *vs.* environment and explores the scientific basis of a possible differential contribution of sire *vs.* dam.

Environmental factors

Both environmental and genetic factors influence development from conception to maturity. Their relative roles can be confusing, particularly for rare or "new" diseases where the critical research to understand the problem has not been accomplished.

Nutrition

To achieve the normal growth and development patterns pre-programmed by a foal's genes requires appropriate nutrients in balanced amounts. Nutritional problems are a recurring source of false reports that identify a "genetic" disease. For example, in the early scientific literature a report on "big head" (osteomalacia) among horses raised in tropical conditions suggested it was a genetic disease of horses. Later, calcium deficiency or calcium/phosphorus imbalance, not a genetic defect, was shown to cause disproportionate bone growth particularly obvious about the head. "Big head" stands as an example of the need to be cautious when aligning defects with their causes. Another problem arises when writers fail to study the literature thoroughly, persistently citing conclusions subsequently shown to be erroneous.

Microorganisms

The organisms responsible for most infectious diseases do not reach the uterus of a pregnant mare. The rare few that do may cause developmental defects or lead to fetal death. The equine herpes virus that causes late gestation abortion is an example of such a problem. If a thorough postmortem examination including a virus assay is not performed, fetal or neonatal deaths in a herd may be incorrectly interpreted as evidence of undefined lethal genes.

Teratogens

A teratogen is an agent that operates during gestation to produce congenital malformations. Appropriate nutrition includes protection from teratogens whose adverse effects on development may be limited to a few days or weeks in early gestation, producing malformations that compromise health or performance throughout life. Teratogens are appropriate to this discussion, because their effects may be falsely attributed to defective genes.

For example, a Paint horse breeder in the southeastern US and his veterinarian tried to understand why the breeder was occasionally getting blind foals of both sexes, but only from certain matings. Since related mares produced the blind foals, although they were not related to the stallion, a genetic problem (possibly maternal) was initially implicated. After extensive searching for an explanation, the most logical interpretation was that the defect had been produced when pregnant mares in pasture grazed certain weeds in early summer (crotonia was the proposed culprit in this case). The teratogenic effect on eye development was apparently critical only during a limited period of gestation. The mares on this farm were foaling on a regular schedule. The same mares would produce affected foals because every year they ate the noxious weed at the time when their fetuses were at the critical gestation stage. Foals from mares bred earlier or later would escape the problem. That related mares were producing the defective foals was by chance. Probably the shared environment was more significant to the problem than shared genes. Since no satisfactory genetic model could explain the situation, the best understanding of the problem implicated a teratogenic effect that a change in pasture management could control.

Maternal vs. paternal environment

It should be obvious that only the mare, not the stallion, contributes to the nutritional and behavioral influences on foal development. The importance of environment relative to the effects of genes, say on racing speed or cutting ability, can be estimated by calculating a heritability index. Few heritability studies in horses have addressed the specific issue of maternal contribution, despite the persistent discussion among horse owners that maternal lineage can influence performance ability.

Nuclear genes

Are allegations of a proportionally larger genetic contribution from the dam than the sire supported by scientific data?

Proportional genetic contributions of autosomal chromosomes

Sire and dam make numerically equivalent contributions of nuclear, non-sex-determining chromosomes (autosomes) to their offspring. In horses, each parent contributes 31 autosomes and one sex chromosome to the foal's 64 chromosome total. The only discussion about unequal contributions of nuclear genes from sire and dam concerns the genes on the X and Y sex chromosomes.

Proportional genetic contributions of sex chromosomes

The X chromosome is the second largest chromosome in the equine karyotype and contains genes critical to maintaining life functions. A female has a pair of X chromosomes. She transmits a set of 31 autosomes and one X chromosome to each of her offspring. The male has a sex chromosome pair whose members are dissimilar in size and genetic information content. The male transmits a set of 31 autosomes and an X chromosome to his daughters and a set of 31 autosomes and a Y to his sons. Male horses and certain infertile females show that a single X chromosome is compatible with life, but no horses (or humans) without at least one X chromosome have ever been found. The Y chromosome is the smallest member of the equine karyotype, containing only a few functional genes, primarily relating to determination of male fertility. Since it is missing in females, the Y clearly is not essential for maintaining life functions of an individual horse, but it is critical for sustaining the species.

For a colt, the autosomal chromosomes come equally from both parents, but his only X chromosome comes from his dam. In all his cells only one allele of an X-linked gene will be present, always of maternal origin. The chromosomes of a filly come equally from both parents. In each female cell only one X is genetically active (serving as a template for protein synthesis). although two copies are available. Either the maternally or paternally derived chromosome in each cell is inactivated, at random, so that the female's body is functionally a patchwork (mosaic) for functioning X-chromosome genes. X-chromosome inactivation in females assures that in each cell only one X-chromosome gene template directs protein synthesis, thus providing compensation for the differences of X-chromosome dosage in colts and fillies.

The case could be made that the dam's X-linked genetic information makes a proportionally smaller contribution in fillies than in colts, contrary to any notion being expressed in the popular press. How substantial is this difference? Using information from mammalian genetics, the difference between fillies and colts for maternal X-linked genes affects fewer than 5% of genes, at maximum. No X-linked genes of positive value for horse performance are yet identified, so at present this differential action is only known to affect situations relating to inherited disease.

Mitochondrial genes

Most genes are located in DNA found in the cell nucleus, but a small amount of genetic material is outside the nucleus in structures called mitochondria. The mitochondrion is a metabolic "power station." Each cell has hundreds to thousands of mitochondria. The genetic information of mitochondria is coded in DNA, but in contrast to nuclear DNA, it is in a single, circular chromosome. Mitochondrial DNA encodes fewer than 40 genes. Thirteen of these genes produce protein subunits that combine with units from nuclear genes to form the active enzymes providing energy for the cell to perform its specific functions. The other genes specify the machinery to translate the encoded protein information.

Perhaps the most important difference from nuclear genes is that mitochondrial genes are inherited from the mother, a **non-Mendelian inheritance** pattern. The egg (maternal) cytoplasm contributes the mitochondria to the zygote. The sperm head delivering the paternal genes to the egg has room only for tightly packaged nuclear chromosomal DNA. Extremely rarely, sperm may contribute mitochondria in addition to nuclear chromosomes. Even in the unusual event that a paternal mitochondrion is transmitted, its genetic influence will be quite overwhelmed by the much greater number of maternal mitochondria.

The genetic contributions of sire and dam are clearly different for genes encoded by mitochondrial DNA. The mitochondrial genes are only a very small part of the entire genetic information package received by the foal from its parents. The percentage of mitochondrial genes relative to nuclear genes, estimated from the best data available, will certainly be smaller than 0.1%.

Diseases associated with mitochondrial genes

The mitochondrial genome is known to be highly conserved during evolution. We can expect that information about mitochondrial genes in other animals will provide useful examples for horses. A few rare human diseases now understood in the light of defects in mitochondrial genes include syndromes of blindness, muscle-nerve weakness and epilepsy. Horse performance traits tied to muscle metabolism controlled by mitochondrial genes might eventually be shown to have a maternal association.

No recombination between mitochondrial genes

Another important difference between mitochondrial and nuclear genes is that the mitochondrial genes do not undergo the recombination processes of sexual reproduction. Differences between mitochondrial sequences within a single maternal line are limited to rare mutational events.

Chromosome imprinting

Studies in mice and humans have shown that a few genes controlling development are exclusively derived from the female parent and others exclusively from the male. Genetic imprinting "turns off" certain genes contributed by the father, so that only the maternal gene is expressed in the offspring, or vice versa. Mutations that disrupt normal imprinting conditions cause rare human diseases, primarily of growth and development. The expression of an imprinted disease-causing gene may depend on whether the problem is transmitted to the child from the mother or father. Transmission patterns of such genes do not follow the classic rules of Mendelian genetics. We have no information about horse genes that may be under this kind of influence but the possibility should be kept in mind. The subtle but distinctive differences in appearance between mules and hinnies may perhaps be explained by imprinting.

Maternal *vs.* paternal genetic contribution

Fillies and colts receive an equal number of nuclear autosomal genes from each parent. For a colt, the dam contributes the only X chromosome, so her contribution to his genome is greater than that of the sire to the extent that the functional genes of the organism are X-linked. Due to X-chromosome inactivation, the dam's contribution of X-linked genes is proportionally less significant to her fillies than to her colts. A foal's mitochondrial genes are contributed with rare exceptions by the dam alone, but overall such genes account for a very small proportion of the functional genes although they are physiologically important. For an unknown number of imprinted genes (currently none known for the horse) the functionally active form is always derived only from one parent, but probably the number of genes participating in this specialized regulation is small and relatively equal between maternal and paternal contribution.

Overall, the comparative genetic contributions of sire and dam are more complicated than discussed in the popular press, but no evidence is available to support a clearly apparent and substantial difference in maternal genetic influence compared with paternal. When we have information about variations in mitochondrial genes of the horse and nuclear genes subject to imprinting, we can better discuss the importance of maternal and paternal contributions relative to specific characteristics.

CHAPTER 16
Genetic descriptions of breeds

Historically, horse breeds were defined by geography (e.g. Arabian, Shire, Exmoor) but in the present-day sense, breeds are sets of related animals sharing selected characteristics and are defined by studbooks. Breeds usually try to maintain a distinguishable phenotype that sets an individual horse of one breed apart from an individual of another. Breeds may also be distinctive for innate abilities such as gait, racing speed or endurance.

Horse owners may get so caught-up in breed promotion that they lose sight of the genetically close relationship of all horses. Genetic tests show allelic frequency differences between breeds, but breeds are not as sharply delineated as different species. The visually distinctive aspects of breed type that allow an individual horse to be assigned to a specific breed may not be reflected by laboratory assays of its genetic markers.

Ideally, a breed description will define the gene pool, not only for pedigree and selected traits like coat color and conformation, but also for genetic markers that are considered neutral with respect to their selective value such as blood group, biochemical and DNA polymorphisms. Allelic frequencies for two highly polymorphic loci *TF* and *VHL20* illustrate both breed differences and similarities (see Chapter 10: Parentage). Such data from several loci can be used to compare genetic diversity statistics between breeds and construct a dendrogram of breed similarities. Readers interested in pursuing population genetics are encouraged to consult a text book such as that by Hedrick (1983).

Calculating allelic frequencies

A simple biallelic locus (albumin) illustrates how allelic frequences are calculated from phenotypic data. Three phenotypes are observed by electrophoresis for the serum protein albumin: ALB-A, -AB and -B. Each ALB-A horse has two *A* alleles; each ALB-AB horse has one *A* and one *B* allele; each ALB-B horse has two *B* alleles.

In a sample of 9848 Paso Finos, there were 1703 horses of type ALB-A, 4799 of type ALB-AB and 3346 of type ALB-B.

- The total number of alleles is 19,696 (twice the number of horses).
- The number of *A* alleles is (2 × 1703) + 4799 = 8205.
- The number of *B* alleles is (2 × 3346) + 4799 = 11,491.
- The frequency of *ALB-A* is 8205/19,696 = 0.42.
- The frequency of *ALB-B* is 11,491/19,696 = 0.58.

Try calculating allelic frequencies for albumin in another breed: among 31,719 Arabians were 6114 horses of type ALB-A, 14,915 of type ALB-AB and 10,690 of type ALB-B. Check your calculations by comparing with the albumin allelic frequencies given for Arabians in Table 11.

	ALB-A	*ALB-B*
Andalusian	0.59	0.41
Arabian	0.43	0.57
Belgian	0.35	0.65
Icelandic	0.43	0.57
Lippizaner	0.08	0.92
Miniature	0.27	0.73
Norwegian Fjord	0.34	0.66
Paso Fino	0.42	0.58
Quarter Horse	0.28	0.72
Shire	0.45	0.55
Standardbred	0.59	0.41
Tennessee WH	0.44	0.56
Thoroughbred	0.17	0.83

Table 11: Both alleles of the serum protein albumin are found in the various breeds of horses. In most breeds, the *ALB-B* allele is the more frequent, but *ALB-A* occurs in such frequency that a substantial number of horses in every breed will be heterozygous. High levels of heterozygosity are usually associated with highly polymorphic, not biallelic, loci. A third allele, *ALB-I*, is rarely seen in a few breeds, but is not included here. The efficacy of this system to detect incorrect parentage among these breeds ranges from 6% to 18%.

ALB data are available for a large number of breeds because the polymorphism is relatively easy to determine by electrophoresis, and historically was the first biochemical polymorphism determined for horses after the technique of starch gel electrophoresis was developed (Stormont & Suzuki 1963).

The Hardy–Weinberg law (H–W)

Under most circumstances, allelic frequencies are not expected to change from generation to generation. This concept may not be intuitively obvious. A popular misconception is that gene frequencies change from one generation to the next depending on gene action. For example, it may be thought that a recessive gene that causes a lethal genetic disease will increase in frequency, probably in analogy to a transmissible, infectious disease. However, only if breeders preferentially select the heterozygous carriers for breeding stock rather than the homozygous unaffected horses will the deleterious gene increase in frequency. Another misconception, contradictory to the recessive scenario just discussed, but probably equally widely believed, is that the dominant (strong) gene will increase in frequency over time and the recessive (weak) will decrease.

Stable allelic frequencies under *random mating* conditions are described by the Hardy–Weinberg law (H–W). For a simple system such as albumin, H–W predicts that the proportion of each homozygous type will be the square of the relevant gene frequency and the proportion of heterozygotes will be twice the product of the relevant gene frequencies. Assigning the symbol p to the allelic frequency of *ALB-A* and the symbol q to the frequency of *ALB-B*, with $p + q = 1$, then the expected phenotypic class sizes in a population of size N are:

Phenotype	ALB-A	ALB-AB	ALB- B
Genotype	*AA*	*AB*	*BB*
Expected number	$p^2 \times N$	$2pq \times N$	$q^2 \times N$

The proportional relationship between phenotypic classes predicted by H–W will be invalid when allelic frequencies are changed by events or forces such as:

- **Migration** of horses into the population (crossbreeding or combining of previously separated populations).
- A gene of positive or negative selective value is linked to the gene being studied (**selection**).
- A population is so small that the influence of chance becomes substantial (**genetic drift**).

In the absence of changes in allelic frequencies by migration, selection, genetic drift or mutation, H–W predicts that phenotypic ratios will remain constant from generation to generation.

Applying H–W to population (breed) data

Applying these formulas to the *ALB* phenotypes for Paso Finos and Arabians given in the previous section, statistical tests show the classes are in H–W equilibrium for Paso Finos (Figure 55) but not for Arabians (Figure 56). Chi-square testing is used to determine whether the data are significantly different from anticipated, but the details of those tests are not provided here.

The Arabian data suggest that the breed in the USA is not randomly mating with respect to albumin. Migration appears to be a plausible cause among the possible reasons for this situation. A large number of Arabians were imported during the 1960–80s and their success in the show ring led to preferential use of them as breeding stock. Particularly, imported stallions were used as mates for the US mares whose pedigree foundations often traced to

Paso Fino	$A = p = 0.42$	$B = q = 0.58$	$N = 9{,}848$
	ALB-A	ALB-AB	ALB-B
Observed	1703	4799	3346
	$p^2 \times N$	$2pq \times N$	$q^2 \times N$
Expected	$0.42 \times 0.42 \times 9848$	$2 \times 0.42 \times 0.58 \times 9848$	$0.58 \times 0.58 \times 9848$
	1709	4787	3352
	Data are consistent with expected H–W proportions		

Figure 55: Phenotypic classes for albumin in 9848 Paso Finos are distributed according to expectations of H–W equilibrium.

Arabian	$A = p = 0.43$	$B = q = 0.57$	$N = 31{,}719$
	ALB-A	ALB-AB	ALB-B
Observed	6114	14915	10690
	$p^2 \times N$	$2pq \times N$	$q^2 \times N$
Expected	$0.43 \times 0.43 \times 31{,}719$	$2 \times 0.43 \times 0.57 \times 31{,}719$	$0.57 \times 0.57 \times 31{,}719$
	5807	15,529	10,383
	Data are ***significantly different*** *from expected H–W proportions*		

Figure 56: Statistical tests of phenotypic classes for albumin in 31,719 Arabians show they are significantly differently distributed from expectations of H–W equilibrium.

different desert sources than those newly arriving. Alternatively, but probably less likely, the skewed classes are due to selection for genes closely linked to *ALB*. Genetic drift due to small population size is clearly not likely in this situation.

For the *ALB* frequencies among breeds in Table 11, only Arabians show the skewed class numbers. The non-random mating in Arabians with respect to *ALB* was not anticipated, but we can predict that the genetic linkage of tobiano pattern with albumin and the phase conservation with *ALB-B* should be evident in data for the Paint breed (Figure 57).

H–W tests of data sets are also important as quality control checks for laboratories performing genetic testing. If markers are regularly misdiagnosed, or if a previously unreported null allele is present, skewed phenotypic class ratios are found compared to those anticipated. In the case of the Arabian data, since the same laboratory (Veterinary Genetics Laboratory, University of California, Davis) performed the tests for the Paso Finos as for the Arabians, the skewed class ratios are not likely to be a technical error. In any case, since these data were generated from samples used for parentage veri-

Paint (overo)	$A=p=0.29$	$B=q=0.71$	$N=5081$
	ALB-A	ALB-AB	ALB-B
Observed	426	2103	2551
	$p^2 \times N$	$2pq \times N$	$q^2 \times N$
Expected	0.29 × 0.29 × 5081	2 × 0.29 × 0.71 × 5081	0.71 × 0.71 × 5081
	430	2096	2554
	Data are consistent with expected H–W proportions		
Paint (tobiano)	$A=p=0.17$	$B=q=0.83$	$N=3390$
	ALB-A	ALB-AB	ALB-B
Observed	40	1061	2289
	$p^2 \times N$	$2pq \times N$	$q^2 \times N$
Expected	0.17 × 0.17 × 3390	2 × 0.17 × 0.83 × 3390	0.83 × 0.83 × 3390
	96	949	2345
	*Data are **significantly different** from expected H–W proportions*		

Figure 57: Tobiano Paint horses have a significant deficiency of phenotypic classes with ALB-A compared with H–W expectations. Overo Paint horses do not show the same class skewing. These differences are anticipated based on the known phase conservation of ALB-B with tobiano.

fication requirements, possible problems with technical errors and null alleles would have been obvious prior to the completion of tests of over 31,000 animals.

Coat color allelic frequencies

Since albumin is not a trait to which horse owners can readily relate, it may be more interesting to compare breeds for allelic frequency differences in coat color genes. For many of the coat color genes one of the pair of allelic alternatives is recessive so the alleles cannot be directly counted from phenotypes. H–W can be used to estimate allelic frequencies in such situations.

Using gray color as an example, we know that all not-gray horses are *gg* and gray horses can be either *GG* or *Gg*. We cannot distinguish by phenotype a *GG* from a *Gg* horse, so we cannot obtain gene frequencies by direct counting. However, from our understanding of H–W, we know that the not-grays are the homozygous recessive class q^2. The square root of their proportion of the population (number of not-gray/number of horses) gives us an estimation of the frequency of *g*. As an example, for Quarter Horses foaled in 1989, the studbook records 6695 grays among 123,294 horses. The frequency of not-grays is 116,599/123,294 = 0.946. The allelic frequency of *g* is the square root of 0.946 which is 0.97. Since $p+q=1$ and $G+g=1$, by subtraction from 1, the frequency of *G* is 0.03.

Since gray horses are infrequent in Quarter Horses, we could also estimate *G* by assuming that all gray horses were heterozygous *Gg*. This assumption would not be correct for every horse, but it provides a reasonable frequency estimation for allelic frequencies in this large sample. In terms of H–W, this means we are assuming that we have no horses in the p^2 class, and all the grays are in the $2pq$ class. If each of the 6695 grays was heterozygous, we would estimate that we have 6695 *G* alleles out of the population total of 246,588 alleles (two times the number of horses in our sample). With this method the frequency of *G* is 0.03, the same as our estimate using the not-grays.

For other loci in which the dominant alleles are infrequent, we can also estimate allelic frequencies by assuming all horses showing the phenotype of the dominant allele, such as C^{cr}, *D* and *RN*, are heterozygous. For the *A* and *E* loci of coat color (the black, bay and chestnut group of basic colors) we must rely on the q^2 class method of H–W to estimate allelic frequencies. The frequency of *e* can be calculated as the square root of the proportion of chestnuts, sorrels, palominos and claybank duns in the data set after removing the grays and roans. The frequency of *a* is calculated as the square root of the proportion of blacks and grullas among the horses with black pigment (blacks, grullas, bays, browns, buckskins and duns).

For the most part, the coat color differences recorded in studbooks are defined by six genes. Comparative coat color allelic frequency data are provided in Table 12 for selected birth years for four breeds, calculated from coat color summaries provided by breed registrars. Students of horse genetics may find it interesting to extend these comparisons to additional breeds.

	Alleles of coat color loci					
Breed (year)	*E*	*A*	*G*	*RN*	*D*	C^{cr}
QH (1989)	0.2	0.7	0.03	0.02	0.03	0.05
MH (1990)	0.5	0.6	0.0002	0	0	0.007
AR (1988)	0.4	0.8	0.02	0*	0	0
PP (1990)	0.3	0.4	0.03	0.01	0.002	0.04

Table 12: Allelic frequencies for the dominant alleles of the major coat color genes. The frequency of the alternative alleles can be obtained by subtraction of these frequencies from one. * Although roan is recorded as a color in the Arabian studbook, it is unlikely to be due to the *RN* gene.

Genetic diversity comparisons

Horses within a pure breed ostensibily are selected to meet defined criteria of ideal breed type. Those knowledgeable about horses can usually correctly assign breed by subjective judgments of distinctive phenotypic traits such as color, body shape and proportion, way-of-going and features of the head and neck. From this phenotypic uniformity it might seem that all horses of a given breed should have a corresponding similarity for traits that can be objectively scored. Yet throughout this book we have provided numerous examples of genetic variation within breeds for blood groups, blood protein and DNA markers. Indeed, despite phenotypic similarity of horses within a breed, the differences provide distinguishing identities for individual horses.

Biological considerations

Survival in the evolutionary sense is thought to be directly related to diversity. A narrow founder base, high levels of inbreeding or a temporary period of decreased population size (bottleneck) can reduce the genetic diversity of a breed. A larger number of founder animals and subsequent minimal levels of inbreeding lead to a broad base of genetic differences from which to select breeding and performance stock. Genetic diversity can be enhanced if a population is subdivided. Typically in wild populations the separation is along geographic lines. In domestic populations, the division may be geographic or may be a product of preferences for performance traits such as shown between trotters and pacers in the American Standardbred by Cothran and colleagues (1987).

Measures of diversity

A simple count of the number of alleles detectable in a given population or breed for a given battery of genetic tests provides one means of comparison. Another measure (average heterozygosity) uses allelic frequencies to calculate the proportion of loci likely to be heterozygous in any individual. Data for 12 horse breeds are provided (Table 13), considering 19 loci of blood typing markers (*A*, *C*, *D*, *K*, *P*, *Q* and *U* of blood groups; *ALB*, *A1B*, *CA*, *CAT*, *ES*, *GC*, *GPI*, *HBA*, *PGD*, *PGM*, *PI* and *TF* of blood proteins).

Breed purity

In the contemporary sense "purity" implies integrity of studbook records. A vital tool to validate the parentage assigned in studbook records is demonstration of genetic marker transmission from parents to offspring. Breed purity in the historical sense is a more difficult concept. Plants that are alleged to be clones of a single progenitor can be tested to validate their uniform identity with other clone members. Animal breeds are genetically quite another story from asexually produced plant cultivars. Genetic diversity within a horse breed is not a sign of "impurity" but is characteristic of sexually reproducing animals and reflects the complex interaction of factors such as founder base size, the subdivision of the breed and the ups and downs of breed size during its history.

Breed	No. of alleles			Calculated average heterozygosity
	Blood groups	Protein	Total	
Thoroughbred	24	34	58	0.295
Andalusian	28	42	60	0.371
Arabian	29	37	68	0.346
Standardbred	29	40	69	0.413
Shire	28	43	71	0.381
Tennessee WH	29	42	71	0.350
Belgian	31	42	73	0.443
Saddlebred	34	47	81	0.386
Peruvian Paso	36	46	82	0.437
Morgan	34	51	85	0.410
Quarter Horse	36	50	86	0.403
Paso Fino	47	58	105	0.433

Table 13: Allelic variation in horses at 19 blood type loci. The theoretical maximum number of alleles known to be detectable was 126. The breeds are sorted in ascending order according to the number of alleles detected. The smallest number of variants was found in Thoroughbreds with a corresponding lowest level of calculated heterozygosity (another measure of genetic diversity, based on allelic frequencies). The greatest amount of genetic diversity was seen in Paso Finos, but essentially equivalent heterozygosity levels are seen in other breeds such as Peruvian Pasos and Belgians (modified from Bowling 1994b).

Breed similarities

Compare the distinctive traits of an American Saddlebred and an American Quarter Horse. It would be unusual to find a random pair of horses, one from each breed, that would be misassigned by phenotype alone to the incorrect studbook. Yet, both breeds trace to early American imports from Europe and certainly share founder animals to some extent. The phenotypic distinctiveness of breeds can appear to provide a persuasive measure of genetic distances. If the distinguishing breed traits are conditioned by only a few genes, it is possible that subjective impressions of the magnitude of breed differences could be quite misleading. Selection programs may have minimal impact on the overall genetic composition, at least within the relatively few generations that separate, for example, Saddlebreds and Quarter Horses.

Comparison of allelic frequencies for various breeds, for example with Nei's distance measurement "*D*," can be used to construct diagrams (dendrograms) of genetic similarity. The dendrogram may reflect the historical and phylogenetic record; however, allelic frequencies and Nei's "*D*" can be affected by mutation, natural selection, artificial selection, founder effect or geographic isolation and can lead to spurious associations if not interpreted with caution.

A dendrogram for 13 horse breeds in the USA (Figure 58) with Nei's "*D*" using allelic frequencies for variants at 21 polymorphic loci provides a picture reasonably consistent with anecdotal evidence of breed relationships.

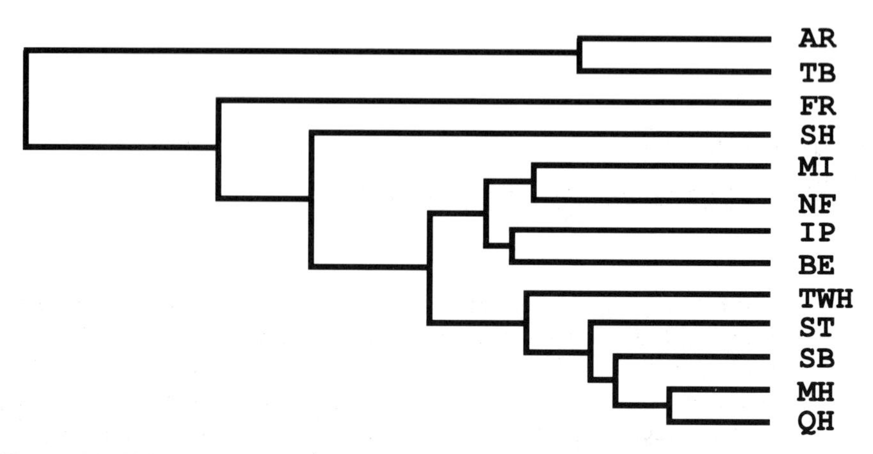

Figure 58: This dendrogram was constructed with Nei's "*D*" using allelic frequencies for variants at 21 polymorphic loci (15 standard loci plus *AP, CAT, CA, C3, GPI, PGM*). The breeds included are Arabian (AR), Belgian (BE), Friesian (FR), Icelandic (IP), American Miniature (MI), Morgan Horse (MH), Norwegian Fjord (NF), American Saddlebred (SB), Shire (SH), American Standardbred (ST), Quarter Horse (QH), Tennessee Walking Horse (TWH) and Thoroughbred (TB) (from Bowling 1992, reprinted with permission of publisher, Elsevier).

The closest similarities are among a cluster of light horse breeds developed in colonial America (Tennessee Walking Horse, Standardbred, American Saddlebred, Morgan and Quarter Horse) that have developed into breeds with separate studbooks only within the last hundred years. Three pony breeds (Norwegian Fjord, Icelandic and Miniature) and Belgians provide a cluster apart from the American breeds and of rather more distant similarities to each other. The clustering of Arabian and Thoroughbred appears to show the influence of "Oriental" stallions in the formation of the Thoroughbred breed. Two draft horse breeds (Friesian and Shire) do not cluster with other breeds of this dendrogram, perhaps an effect of their different origins or of population bottlenecks affecting allelic frequencies.

In the future, we can look forward to breed similarities compared through analysis of three types of genetic information: coding genes (blood groups and protein polymorphisms), non-coding genes (microsatellite polymorphisms) and mitochondrial sequence polymorphisms. This composite view of the genetic material of horse breeds has the potential to be a powerful tool to corroborate or modify anecdotal origins of horse breeds.

CHAPTER 17
The horse gene map

Gene mapping for horses lags in contrast to other livestock species and humans, but relatively rapid progress is expected by relying heavily on techniques and maps developed for other mammals (O'Brien *et al.* 1993). Genes that can be put on the map include those describing physical traits (e.g. coat color), enzymes or other proteins (e.g. transferrin), diseases (e.g. HYPP), or DNA sequences of unknown function (e.g. microsatellites). The estimated 50,000–100,000 genes of the horse will be arranged over 32 pairs of chromosomes—31 autosomes plus the X and Y sex chromosomes. DNA sequence and gene linkage relationships among genes highly conserved in mammals will provide a productive starting point for the horse map.

For humans, over 4000 genes are recognized, and chromosomal loci are known for more than 1300 genes distributed over all the chromosomes (McKusick 1990), with active research rapidly adding to these numbers. For horses, 24 genes at least have tentative autosomal chromosomal assignments involving eight of 31 autosomes (Table 14).

A gene map of the horse could provide horse breeders with genetic tools to select disease-free, high performance animals. Linked genes are currently used by tobiano breeders as an *indirect test* to identify horses likely to be homozygous for the spotting pattern. Difficulties associated with linkage tests (uninformative matings, the rare possibility of misdiagnosis due to undetected recombination, and the need to test parents and other offspring as well as the horse in question) may obviate their usefulness. The most important role for gene linkage maps is to identify likely candidate genes described from other mammals that may point the way to a *direct test* for the gene in horses. In the case of tobiano, spotting genes in the mouse linked to the same genes as in the horse may lead to a test for the tobiano gene that avoids the problems associated with linkage testing.

A modest goal for the horse gene map is to have 350–400 reference genetic markers—evenly spaced throughout the chromosomes—so that any trait of interest has a good chance of being close to a defined group of genes that can be related to the maps of human and mouse. Probably 1000 genetic markers need to be characterized to meet the goal.

Establishment of linkage groups (LGs) is achieved by:

- Tracking the concordant presence or absence of trait pairs within family groups.
- Assaying for the concordant presence of gene markers in somatic cell hybrid lines.
- Physically assigning genes to the same chromosome with *in situ* hybridization.

Linkage mapping from family study

The classic research technique for identifying a pair of chromosomally linked genes is to compare the numbers of offspring that show trait pairs like the parents with those that have a new combination, not present in either parent. Data from a large number of offspring are required for significant results. The proportion of trait recombinants among the offspring is a measure of the chromosomal distance between the traits. Genetic map units are in centimorgans (cM), named after the geneticist Thomas Hunt Morgan. Two traits are about 1 cM apart if they show recombination 1% of the time. The physical distance between trait loci separated by a genetic distance of 1 cM is roughly one million base pairs of nucleotide sequence.

An example of what linkage data might look like to a horse breeder is provided by a coat color example from Sponenberg and colleagues (1984) involving roan, bay and chestnut. A bay roan Belgian stallion bred to eight chestnut mares sired 30 bay roans, 25 chestnuts, 1 bay and 1 chestnut roan. The stallion is clearly heterozygous for both extension and roan (*Ee RNrn*). If these genes were on different chromosomes we would anticipate equal numbers (about 14 foals) in each of the four phenotypic classes. The very skewed distribution seen in these data, with the numbers of progeny in parental color classes (bay roan and chestnut) exceeding the numbers in classes with new color combinations (bay and chestnut roan), is typical for linked genes. In this case, the recombinant animals (2 of 57) amount to 3.5% of the offspring, for an estimated map distance of 3.5 cM.

Not all traits and not all families are informative for linkage analysis. Linkage mapping with family study is best suited to animals with a short gestation period and large litter size, and so has not been a highly productive mapping technique for horses. Currently six autosomal linkage groups (LGs), numbered in order of their discovery, have been defined for the horse using family segregation data. LGs II, III and IV include clusters of genes highly conserved among mammals (e.g. *ALB* and *GC*; *MHC* complex; and *GPI*, *CRC* and *A1B*, respectively). LG II currently has more genes than the others.

X-chromosome assignments are highly conserved among all mammalian species. Those so far determined for the horse predictably match those for other species, although the linear order may be rearranged.

Somatic cell hybrid mapping

This technique identifies gene synteny (genes on the same chromosome) by scoring the correlated presence of trait markers in clones of horse and mouse cell hybrids. To create the hybrids, horse and mouse cells growing in tissue culture are fused. The chromosome mixture of the fusion cell is unstable and some horse chromosomes are lost. After a few generations in culture, chromosome numbers in the hybrid cells stabilize and a panel of clones can be established, containing one or a few horse chromosomes in different combinations. This technique will not identify physical traits like coat color, but can be used for enzyme, microsatellite and other DNA markers. Somatic cell hybrids do not readily allow genetic distance measurements, but provide gene mapping information without the expensive and time consuming need to breed horses. Williams and colleagues (1993) identified two syntenic groups (*LDHB*, *PEPB* and *IGF1*; *NP*, *MPI* and *IDH2*) and possibly a third (*ADA* and *PEPC*), but they do not yet have LG or chromosomal assignments. Bailey and colleagues (1995) identified another six syntenic groups based on microsatellite markers (KS1: *ELA-DRA*, *TNFA*, *HMS5* and *HTG5*; KS2: *HTG3* and *HTG13*; KS3: *HTG4*, *HTG8* and *HMS3*; KS4: *HTG6* and *HMS1*; KS5: *HTG7*, *HTG9* and *HMS6*; and possibly KS6: *HTG12*, *HMS7* and *ECA2*). Research in progress is expected to identify syntenic relationships among several hundred DNA markers.

In situ mapping

Genes may be mapped to specific chromosomes by *in situ* techniques that treat dividing cells with specific gene probes, labelled radioactively or fluorescently, followed by karyotyping to locate the chromosome binding sites. The current method of choice for physical mapping of genes to chromosomes is FISH (fluorescence *in situ* hybridization). Taking advantage of conserved gene homologies among mammals, human and pig gene probes have been used successfully to locate horse genes to chromosomes. Eight autosomes have been assigned specific genes using this technique (Table 14), placing three of the nine linkage groups (LG I, LG III and LG IV). This is another area for which research in progress is expected to significantly increase our understanding of horse genetics.

Genetic relationships among autosomal genes of horses

Gene	Symbol	LG	Tech	Chr	References
A (blood group)	*A*	III	F	(20q)	Bailey *et al.* 1979
Adenosine deaminase	*ADA*	U	S	nd	Williams *et al.* 1993
A-1-B glycoprotein	*A1B*	IV	F	(10)	Andersson *et al.* 1983a, Weitkamp *et al.* 1982
Albumin	*ALB*	II	F	nd	Sandberg & Juneja 1978, Trommershausen-Smith 1978

Gene	Symbol	LG	Tech	Chr	References
Calcium release channel	*CRC*	IV	I	10pter	Chowdhary *et al.* 1992
Complement component 3	*C3*	U	I	7p	Millon *et al.* 1993
Complement component 4	*C4*	III	F	(20q)	Kay *et al.* 1987a
Carboxylesterase	*ES*	II	F	nd	Andersson & Sandberg 1982
DRA (major histocompatibility complex)	*DRA*	KS1	S	(20)	Bailey *et al.* 1995
ECA2 (microsatellite)	*ECA2*	KS6	I	1q2.1	Sakagami *et al.* 1995
ECA3 (microsatellite)	*ECA3*	I	I	2p1.3–4	Tozaki *et al.* 1995
Equine soluble class I antigen	*ESCI*	III	F	(20q)	Lew *et al.* 1986
Glucose phosphate isomerase	*GPI*	IV	F, I	10pter	Andersson *et al.* 1983a, Harbitz *et al.* 1990
Group-specific component	*GC*	II	F	nd	Sandberg & Juneja 1978
Glutamate oxaloacetate transaminase—mitochondrial	(GOT_m)	II	F	nd	Andersson *et al.* 1983b
Extension	*E*	II	F	nd	Andersson & Sandberg 1982
F13A	*F13A*	III	F	(20q)	Weitkamp *et al.* 1989
Hemoglobin-alpha	*HBA*	U	I	13q	Oakenfull *et al.* 1993
Haptoglobin	*HP*	II	F	nd	Weitkamp *et al.* 1985
HMS1 (microsatellite)	*HMS1*	KS4	S	nd	Bailey *et al.* 1995
HMS3 (microsatellite)	*HMS3*	KS3	S	nd	Bailey *et al.* 1995

Gene	Symbol	LG	Tech	Chr	References
HMS5 (microsatellite)	*HMS5*	KS1	S	(20)	Bailey *et al.* 1995
HMS6 (microsatellite)	*HMS6*	KS5	S	nd	Bailey *et al.* 1995
HMS7 (microsatellite)	*HMS7*	KS6	S	(1)	Bailey *et al.* 1995
HTG2 (microsatellite)	*HTG2*	II	F	nd	Marklund *et al.* 1994
HTG3 (microsatellite)	*HTG3*	KS2	S	nd	Bailey *et al.* 1995
HTG4 (microsatellite)	*HTG4*	KS3	S	nd	Bailey *et al.* 1995
HTG5 (microsatellite)	*HTG5*	KS1	S	(20)	Bailey *et al.* 1995
HTG6 (microsatellite)	*HTG6*	KS4	S	nd	Bailey *et al.* 1995
HTG7 (microsatellite)	*HGT7*	VI, KS5	F	nd	Marklund *et al.* 1994, Bailey *et al.* 1995
HTG8 (microsatellite)	*HTG8*	KS3	S	nd	Bailey *et al.* 1995
HTG9 (microsatellite)	*HTG9*	KS5	S	nd	Bailey *et al.* 1995
HTG12 (microsatellite)	*HTG12*	VI, KS6	F, S	nd	Marklund *et al.* 1994, Bailey *et al.* 1995
HTG13 (microsatellite)	*HTG13*	KS2	S	nd	Bailey *et al.* 1995
Insulin-like growth factor-1	*IGF1*	U	S	nd	Williams *et al.* 1993
Isocitrate dehydrogenase 2	*IDH2*	U	S	nd	Williams *et al.* 1993
K (blood group)	*K*	I	F	(2p)	Sandberg 1974a
Lactate dehydrogenase B	*LDHB*	U	S	nd	Williams *et al.* 1993
Major histocompatibility complex	*MHC* (*ELA*)	III	F, I	20q	Bailey *et al.* 1979, Bernoco *et al.* 1987a, Ansari *et al.* 1988, Mäkinen *et al.* 1989

Gene	Symbol	LG	Tech	Chr	References
Malic enzyme	*ME1*	IV	F	(10pter)	Weitkamp *et al.* 1982
Mannose phosphate isomerase	*MPI*	U	S	nd	Williams *et al.* 1993
Nucleolar organizer region	*NOR*	U	I	1p, 28, 31	Kopp *et al.* 1981, 1988, Millon *et al.* 1993
Nucleoside phosphorylase	*NP*	U	S	nd	Williams *et al.* 1993
Peptidase B	*PEPB*	U	S	nd	Williams *et al.* 1993
Peptidase C	*PEPC*	U	S	nd	Williams *et al.* 1993
Phosphogluconate dehydrogenase	*PGD*	I	F, I	2p	Sandberg 1974a, Gu *et al.* 1992
Protease inhibitor	*PI*	V	F	nd	Bowling 1986
Roan	*RN*	II	F	nd	Andersson & Sandberg 1982, Sponenberg *et al.* 1984
Steroid 21-hydroxylase	*21OH*	III	F	(20q)	Kay *et al.* 1987a
Tcp-1	*TCP1*	U	F	nd	Langemeier *et al.* 1993
Tcp-10	*TCP10*	U	F	nd	Langemeier *et al.* 1993
Tobiano	*TO*	II	F	nd	Trommershausen-Smith 1978
Tumor necrosis factor A	*TNFA*	KS1	S	(20)	Bailey *et al.* 1995
U (blood group)	*U*	V	F	nd	Bowling 1986

Table 14: Autosomal linkage and chromosomal assignments for the horse. In the linkage group (LG) column, "U" means unassigned and KS numbers are syntenic assignments from Bailey *et al.* 1995. In the techniques (Tech) column, "I" is for *in situ*, "F" for family study, "S" for somatic cell hybrid. In the chromosome (Chr) column, the number designates the chromosome, a "p" puts the marker on the short arms of a metacentric, a "q" on the long arms, and "pter" on the end of the long arm, "nd": not determined, and numbers in parentheses are indirect assignments through linkage or synteny to a marker located with *in situ* testing. Estimated map distances determined from family studies can be obtained from the references cited. Current as of July 1995.

CHAPTER 18
Equus

In the Linnean system of organizing the living world, equines are classified as follows:

Kingdom: Animal
Phylum: Vertebrate
Class: Mammal
Order: Perissodactyl
Family: Equid
Genus: Equus

The perissodactyls are herbiverous, hoofed mammals with an odd toe number. The closest relatives to the equines are the other perissodactyls—rhinoceros and tapir. For a more detailed discussion of the family consult Groves (1974) and Clutton-Brock (1992).

Chromosomal differences among the equine species

The fossil record shows that the extensive evolution of equines occurred in North America, but the living wild species today are found only in Africa and Asia. Chromosomal morphology and numbers in the living members of the genus *Equus* present a heterogeneous picture. As an illustration of this diversity, the chromosome numbers for the species and some subspecies are shown in Table 15.

Numerical as well as morphological chromosomal differences show that the speciation process in equines involved exchanges of DNA sequence blocks in many chromosomes. Based on the fossil record, these changes occurred over a relatively short, well-documented time (Benirschke & Ryder 1985). Such extensive rearrangements are not typically found in other genera, for example among felines (cats) and bovines (cattle and buffalo). Despite the array of numerical and structural differences, several prominent chromosomal similarities are found, particularly within the horses, the asses and the half-asses (hemiones). The zebra species show significant differences from each other as well as the other species.

Species	Common name	Chromosomes
E. przewalskii	Przewalski's wild horse (takh)	**66**
E. caballus	domestic horse	**64**, 65
E. africanus somaliensis	Somali wild ass	**62**, 63
E. asinus	domestic ass (donkey)	**62**, 63
E. hemionus hemionus	Mongolian wild ass (dziggetai)	?
E. hemionus onager	Persian wild ass (onager)	55, **56**
E. hemionus kulan	Transcaspian wild ass (kulan)	**54**, 55
E. hemionus khur	Indian wild ass (khur)	?
E. kiang	Tibetan wild ass (kiang)	51, **52**
E. grevyi	Grevy's zebra	**46**
E. burchelli boehmi	Grant's zebra (common zebra)	**44**, 45
E. zebra hartmannae	Hartmann's mountain zebra	**32**

Table 15: The diploid chromosome number for each equine species is indicated in boldface. In several species Robertsonian polymorphisms (fusion–fission systems between the arms of a metacentric and two acrocentric elements) have been described and are shown in regular typeface. Data are compiled from information in Ryder *et al.* 1978, Benirschke & Ryder 1985, Bowling & Millon 1988 and Ryder & Chemnick 1990.

Przewalski's horse and the domestic horse

Przewalski's horse has extensive chromosome homology with the domestic horse. Two acrocentric pairs in *E. przewalskii* appear to be combined in one metacentric chromosome (chromosome 5) in the *E. caballus* karyotype. A Robertsonian translocation event can explain the difference between the two species; either fusion or fission is possible, assuming one species is derived from the other, depending on which arrangement is ancestral.

Genetic homology between Przewalski's horse and the domestic horse is also evident in similarity of genetic markers. Among systems in the conventional battery of blood typing tests, most of the markers recognized for Przewalski's horse are also found in domestic horses. Markers apparently unique to Przewalski's horses occur in TF, XK, ES and PI systems, but are not present in all animals. One genetic study (Bowling & Ryder 1987) suggested that the similarities imply a subspecies rather than a full species separation. Although a domestic mare is known to be included in the pedigrees of many of the living Przewalski's horses recorded in the studbook, it has not been possible to trace specific markers that she might have introduced. For example, among seven TF alleles recognized in Przewalski's horse, only one is unique to Przewalski's horses. Among the six TF specificities shared between domestic and Przewalski's horses, at most two alleles could have been contributed by the domestic mare, but so far they have not been defined. Studies are continuing on genetic markers in Przewalski's horses, including DNA markers, so eventually the relationship between these two taxa may be better resolved.

All breeds within the domestic horse species have the same diploid chromosome number (2n = 64), with the exception that 2n = 65 has been reported for some Caspian Ponies (Hatami-Monazah & Pandit 1979). The extra chromosome could be explained as centric fission of a metacentric chromosome into two acrocentrics, although the authors suggested that the origin followed natural hybridization between domestic horses and Przewalski's horse.

Somali wild ass and domestic ass

The ass karyotype numerically differs from that of the horse by a single chromosome pair, but the morphology and banding patterns of the individual chromosomes are quite different from those of the horse. The karyotypes of the Somali wild ass and domestic ass cannot be distinguished. A Robertsonian polymorphism has been described in each of these species, probably involving the same large metacentric chromosome (Benirschke & Ryder 1985, Bowling & Millon 1988).

Asiatic wild asses

The taxonomy of the Asiatic wild asses has a considerable degree of uncertainty. They have been difficult to study because of their remote locations in Asia and scarcity in zoo collections and in the wild. The kulan and onager (hemione) karyotypes are similar, but differ with extensive rearrangements from African wild asses. Kiangs have from two to four fewer chromosomes than the hemiones. Reports of population polymorphisms for kulan, onager and kiang involve Robertsonian polymorphisms, probably of the same homologous elements (Ryder & Chemnick 1990).

Zebras

The best-known feature of the African zebras is probably the presence of extensive, distinctive striping patterns. Grevy's zebra is the largest and northernmost of the zebras, and has the largest number of chromosomes. The common zebra differs numerically by a single chromosome pair from Grevy's zebra, but chromosome morphologies show several rearrangements. The third species, the Hartmann's mountain zebra, is physically the smallest and southernmost and has the lowest chromosome number. As might be anticipated from the discussion of the other species, a Robertsonian rearrangement has been identified for the zebra, in a stallion at the London Zoological Society (Whitehouse *et al.* 1984).

Species hybrids

Despite extensive chromosomal differences, the various equine species generally can be successfully crossed to produce viable progeny (Figure 59). Such hybrids show that despite chromosomal changes the overall genetic content between members within this genus is not markedly different. Except for hybrids between Przewalski's horse and the domestic horse or among the hemiones, the hybrid offspring are invariably sterile.

Figure 59: Species hybrid between a Chapman's (common) zebra male and a bay domestic mare at Askania Nova, Ukraine. The hybrid is bay like its dam, but with extensive striping.

Rare fertility in mules and hinnies

Historically, the species cross between domestic horse and donkey has been agriculturally highly important. The offspring of a female horse and a male donkey is called a mule; the reciprocal cross between a female donkey and a male horse produces a hybrid said to be of slightly different appearance called a hinny. The mule is said to possess an ass-like head with long ears and an ass-like tail. The hinny has a finer head—more horse-like and with shorter ears than a mule—and a horse-like tail. The donkey differs in chromosome number from the horse by one fewer pair, but significant differences in morphology imply multiple rearrangements have occurred during evolution so that only extremely rarely would hybrids be expected to produce genetically balanced gametes during meiosis. The ovaries of most female horse-/donkey hybrids are atrophic and males are azoospermic. The rare occurrence of fertile female hybrids has been rumored, but only recently verified scientifically. Ryder and colleagues (1985) used karyotyping and blood typing to confirm the parentage of a jack (male) mule foal by a Welsh pony stallion out of a molly (female) mule. Karyotyping studies of fertile female mules and hinnies in China have also been reported (Rong *et al.* 1988). The lack of evidence for male fertility may reflect lack of opportunity, but is more likely to be an example of Haldane's Law, which predicts male sterility in species hybrids.

Endangered species

At one time widely distributed over Europe and Asia, Przewalski's horse has become extinct in its last wild range in Mongolia during the mid-twentieth century. The species is preserved in zoological collections throughout the world and attempts are being made to reintroduce it back into the wild in Mongolia. Currently zoological collections number about 1000 animals. A studbook record traces these animals to 12 wild-caught horses (four mares and eight stallions), plus a domestic Mongolian horse introduced as a nurse mare for a wild-caught suckling foal.

The African wild asses are extremely threatened in the wild and few examples are found in zoological collections. Asiatic wild asses are difficult to maintain in zoo conditions, but several species are severely threatened in their native habitat and may require significant conservation measures to survive.

Likewise for some zebra species, significant conservation programs are needed. The living species are present both in the wild and in zoos, but only the common zebra has sufficient habitat and numbers for long-term survival unassisted by conservation management.

Feral species

The wild species have been extinct in North America for thousands of years (except for zoo collections), but feral equines are wild and free-ranging in several locations, primarily in the western USA under the jurisdiction of government agencies (Figure 60). The feral equines are representatives of both

Figure 60: Feral horses in winter in southeastern Oregon.

domestic equine species, horses and asses. Generally, the species occupy different ecological niches, with the asses located on the more arid ranges. Species hybrids (mules) occur among feral horse herds where the ranges overlap.

The origins of the feral animals in most American regions are largely undocumented, but anecdotally trace to the earliest importations of domestic equines by the Spanish explorers and early colonizers. Studies show substantial genetic diversity within herds and some variants rare in recorded breeds are in high frequencies in feral herds (Bowling 1994b). These feral equines provide a gene pool naturally selected for combinations maximizing soundness, hardiness and thriftiness.

CHAPTER 19
Frequently asked questions

As an alternative format for presenting information, here are answers for specific situations that are representative of frequently asked questations about horse genetics. You will have no difficulty sensing my frustration that good questions often have inadequate answers because basic genetic research data for the horse are lacking.

Color, hair characteristics and gait

Blue eyes?

Q: I have a Welsh Pony mare that just foaled her third foal, all by the same stallion. Much to my surprise one of the filly's eyes is partially blue. Neither the sire nor the dam has a blue or partial blue eye. None of the other seventeen foals by this stallion has had this trait nor have the mare's other two foals. How is this trait inherited?

A: Unfortunately, the scientific literature does not provide a definitive scheme for the inheritance of blue or partial blue eyes (*heterochromia irides*) in the horse. Blue eyes are a frequent feature in breeds with major spotting genes (tobiano, overo, appaloosa). Blue eyes are expected in the extreme color dilution phenotypes (cremello and perlino) where their sun sensitivity may be a significantly undesirable characteristic. Blue-eyed foals are not anticipated in breedings between non-spotted, non-dilute parents but certainly occur in many breeds, although rarely. Blue or partial blue eyes in non-dilute colored horses do not seem to be sun sensitive or otherwise defective.

Inheritance does not seem to follow a simple Mendelian pattern. In most cases neither parent of a blue-eyed foal has blue eyes, so it is not a simple dominant trait. Breeding data from blue-eyed stallions bred to brown-eyed and blue-eyed mares might be quite useful to help sort out the inheritance. These data may be difficult to obtain since the common notion is that a solid-color horse with blue eyes is undesirable and seldom would mare owners be enticed to breed to such a stallion even if he was otherwise outstanding. The inheritance of blue eyes is but one of the unanswered genetic dilemmas in breeding horses. Perhaps a breeder has some data that could be shared or would be willing to underwrite this project.

Tricolored Pinto?

Q: My neighbor has a stallion she advertises as a tricolored Pinto. This horse looks like a bay tobiano to me. She claims his color is very special. I can't find any discussion of this color in horse genetics books. Can you tell me more about this color?

A: The advertisments that I have seen for "tricolored" tobianos look like either bay or buckskin tobianos. I am not aware of any special genetic situation that would give rise to these color combinations. Perhaps the point owners are trying to make is that everyone else is keen to have a black tobiano and their horse provides an alternative.

Coat color genes?

Q: What kind of blood samples do I need to send to get a readout of the coat color genes of my buckskin Peruvian Paso mare?

A: The symbols on the blood type marker printouts are not codes for coat color genes. Assignment of coat color genotypes comes not from laboratory testing but from study of the color of the horse, its offspring and parents. Your mare will be: *ww, gg, E–, A–,* $C^{cr}C$*, dd, zz, toto, oo, lplp, rnrn*.

Roan lethal?

Q: I have a bay roan Quarter Horse stallion that is a many times champion cutting horse. Some people have told me I should not stand him at stud because roan is a lethal gene. He is a wonderful horse. I would like to have lots more like him and I will breed as many mares of my own as I can afford to keep. What is all this talk of roan as a lethal gene—my stallion is clearly alive.

A: Studbook research suggests that the allele for roan color is a homozygous lethal. A horse with only one copy of the allele—such as your stallion—obviously has no problems from having the gene. Horses with two copies of roan, such as could occur from breeding a roan mare to a roan stallion, allegedly die as embryos. Unfortunately we have no direct data from breeding trials to verify this embryo lethality process.

With your interest in this color, you may be able to provide missing research data. What we need is for you to acquire some bay roan mares to breed to your stallion. Not all bay roan mares will do. They must be bred like your stallion, with one bay parent and one chestnut roan or sorrel roan parent. Since you will be pasture breeding these mares, we will not be able to follow the possibility of early embryonic loss but we can look at color classes to provide another kind of evidence. We will be using the known linkage of roan to extension to evaluate the hypothesis of lethality for homozygous roan. With this cross, three classes of offspring colors are anticipated: bay, bay roan and chestnut roan.

Genetic contribution from mares	Genetic contribution from stallions	
	E rn	*e RN*
E rn	*EE rnrn* bay	*Ee RNrn* bay roan
e RN	*Ee RNrn* bay roan	*ee RNRN* chestnut roan

If roan is *not* a homozygous lethal you will get about 25% bay, 50% bay roan and 25% chestnut roan foals. If roan *is* a homozygous lethal your mares will not be producing chestnut roans, except as rare heterozygous roan recombinants between the linked genes (expected for about one foal in 50). To evaluate the ratios as fulfilling one hypothesis or the other, we will need your mares to produce at least 20–30 foals. Now the advantages of using fruit flies and mice for genetic research become magnificently obvious! Keep in touch. Maybe other owners would also like to participate in this project since we could combine data from several stallions as long as the specific genotype requirements are met.

Lethal white in mules?

Q: I have long been an admirer of mules. I have an overo Pinto mare that I would like to breed to my neighbor's spotted jack. Do we need to be concerned about the possibility of producing a lethal white mule?

A: What an interesting question! Unfortunately, we have no good answer. The spotting pattern in asses, apparently inherited as a dominant trait, is probably homologous to one of the spotting patterns in horses, but we are not aware of any data to provide specific details. Since you are clearly aware of the possibility of a lethal white foal, if you can emotionally and economically afford it, please make the cross (several times) and let us know what happens.

"True" black?

Q: I have a black Arabian stallion with a lot of roaning in his flanks and on the top of his tail. The owner of another black stallion says my stallion is not a true black and it is improper or even unethical of me to advertise him as black. If I send you a photograph will you write a letter stating that he is a true black so that I can advertise him as such?

A: I don't know a genetic definition for "true" black. Two genetic situations are proposed to account for black color, one a dominantly inherited pattern, the other a recessive. I don't see any compelling reason to label one as true and the other as false. Probably the dominant black type doesn't occur in Arabians, in any case. The roaning trait shown by your stallion is best likened to a marking. Does your nemesis also reason that a true black would have no white markings whatsoever? The problem you describe is one of the unfortunately too common examples of breeders invoking pseudoscience to justify their personal preferences. Stand up for logic and reason, show your stallion as a performance star and advertise him as a desirable horse regardless of his color.

Curly coat?

Q: I am the proud owner of a mustang. This special mare has a very curly hair coat. Her mane is like corkscrews. Is this an inherited trait? If yes, to what should I breed her to get a curly foal?

A: Occasional horses with curly coats are seen in various breeds in the USA, such as Percheron (Blakeslee *et al.* 1943), Quarter Horse, Missouri Fox Trotter and among feral horses. Studbook records of the American Bashkir Curly Registry suggest both dominantly and recessively inherited types of curly coat may be found (Sponenberg 1990). In my experience the dominantly inherited type of curliness is the one found in feral horses. You can probably expect your mare is heterozygous for the curly trait and that 50% of her foals will have a curly coat regardless of the coat type of the stallion you use.

Champagne color?

Q: This spring I bred my black Tennessee Walking Horse mare to a stallion whose rare and beautiful color the owner calls "champagne." What are my chances for getting a foal of this color?

A: The color term "champagne" is specifically applied to a black-diluted color in Tennessee Walking Horses. The gene responsible also affects bays and probably chestnuts, but the names given those combination colors are usually the conventional terms buckskin and palomino (or the more vague designations dun and yellow). Champagne foals are born a smokey gray with blue eyes and pinkish-gray skin. As the foals age, the eyes darken to hazel or brown, but the skin and coat remain light and distinctive. The eyes of newborn foals of the bay-diluted combination are also blue and darken with age, but information is not available about the eye color of the red (pheomelanin) foals.

The color trait is inherited as a dominant. It resembles the palomino/buckskin dilution effect but is probably due to a different gene. From its phenotype it does not seem to be an allele of dun or silver dapple, but those possibilities and their combinations with this gene remain to be identified. We also have no information about the color of homozygotes. Champagne may be the same color as *globrunn* in Icelandic horses and lilac dun in other ponies.

Breeding data for the stallion you used show that he is heterozygous for the dilution and homozygous for black pigment (*EE*). You probably have equal chances for either a champagne or a black foal.

Inheritance of ambling gait in Arabians?

Q: My Arabian gelding is 20 years old this year. He has been a wonderful trail companion for 15 years and introduced me to the pleasures of what I call an ambling gait. He also can do the basic walk, trot and canter repertoire, but we both prefer his four beat amble for our pleasure rides. Am I going to be able to find another gaited Arabian to replace him or do I have to switch to a breed that has been traditionally selected for this trait?

A: Gaited Arab horses are reported from time to time but are certainly not common. One of the current top endurance horses is an Arabian with an amble or running walk, but we know of no one who is specifically breeding Arabian horses for this trait. We are not aware of controlled breeding trials that would define the trait inheritance pattern in Arabians. If you put a "horse wanted" ad in a national Arabian publication you will probably find several gaited horses for sale, but you are unlikely to have a large number to choose from. If you do not want to travel very far to try out a horse, you might have to switch to another breed so you have several local examples and a wider choice of color, age and sex.

Medical problems and possible genetic disorders

Screening for defective genes?

Q: What kind of blood samples do I need to send to identify the defective genes of a yearling Quarter Horse gelding I am thinking of buying? I am particularly concerned about defects of feet and legs. My first mare went lame from navicular and I don't want a repeat of the problems I had with her.

A: At this time the only defective gene that can be diagnosed by DNA testing is the dominantly inherited muscle paralysis disease HYPP. No genes controlling conformational traits have been identified. Numerous examples can be found of eminently successful performance horses with less than perfect conformation, but you are certainly justified to identify areas of special importance for your program. You may especially want to avoid horses that have been retired to broodstock status because of a problem with navicular disease (although they need not have offspring with this problem). Conformation that predisposes to unsoundness may be difficult to spot in a young horse, so be sure to try to see the parents and ask questions about how they held up to performance trials. Good luck!

Parrot mouth?

Q: Six months ago I bought a very expensive yearling Paint colt as a stallion prospect with the guarantee that I could return him if he was genetically defective. He is starting to develop a parrot mouth and I want to send him back, but the seller says I must prove this is a genetic problem. Can you send me a signed statement that this is a genetic problem so I can get my money back?

A: Although defective genes may effect patterns of abnormal bone growth, the inheritance of parrot mouth (brachygnathism) has not been defined. Gaughan & DeBowes (1993) report a brachygnathic sire with one affected foal out of 41, so inheritance is clearly not due to a dominant gene. Males are more often affected than females, another fact pointing inheritance away from a conclusion of a simple Mendelian gene. Probably parrot mouth is a polygenic trait that can be influenced by environmental components (nutrition) as well as by genes. No laboratory tests are available to define the relative importance of each component in particular cases. In your situation, the genetic component could probably only be proven by breeding the colt to normal and affected mares to see whether and how frequently he transmitted the trait. I agree that this suggestion is not a practical solution to your problem, but at some point these studies need to be done and results published so that sellers and buyers can be the responsible horse breeders that they want to be. Your situation emphasizes the importance of listing specific conditions in the contract that would allow you to return the colt for a refund of purchase price. Buyer protection laws in your state may help you even if you did not specify conditions in the contract. However, in most animal transactions probably the assumption is "buyer beware." The buyer who agrees to assume unnamed risks (no guaranteed refund) may be successful in negotiating a lower price for a young, but unproven breeding prospect.

Cryptorchidism?

Q: I am a small breeder with only two Tennessee Walking Horse mares and no stallion. Every year I breed them to world champions. Last year both mares had colts by the same stallion. They are nice colts, but one is a cryptorchid as a yearling. My veterinarian cannot palpate his second testicle. I want you to help me sue the stallion owner to reimburse me for my stud fees and all my expenses in raising these foals and to prevent him from defrauding other mare owners. He says that my colt is the only cryptorchid his stallion has sired from nearly 100 foals. My mares have each had one other colt by another stallion, but neither was a cryptorchid.

A: Cryptorchidism is the failure of one or both testicles to descend into the scrotum. If the retained testicle is abdominal, then it is unlikely to descend, but a testicle in the inguinal canal of a young colt may later descend. Cryptorchidism is another one of those assumed genetic traits with insufficient evidence to support the popular assumption. The problem seems to occur in all breeds. It is probably an elusive polygenic trait with a threshold effect and the extra complication that trait expression is limited to males. Possibly some cases have an environmental cause (such as poor nutrition or trauma at birth or later). If this is a genetic trait, most likely your mare as well as the stallion contributed problem genes to the colt. The standard breeding contract you signed only guaranteed a foal that could stand to nurse. Perhaps you should have discussed a different set of guarantees with the stallion owner prior to breeding your mares.

Inheritance of HYPP?

Q: I have a seven-year-old Appaloosa mare that has tested positive for HYPP, but she has no symptoms of the disease that I have ever seen. Can she still transmit the gene if I breed her to a stallion that is negative for the HYPP gene?

A: Gene positive HYPP horses (N/H) that do not appear to have muscle paralysis episodes can transmit the defective gene. Minimally affected heterozygous HYPP horses may have less of the abnormal gene product than more severely affected heterozygotes (Zhou *et al.* 1994) but we have no evidence for variability in trait transmission between minimally and severely affected heterozygotes. Diet and exercise managed HYPP gene positive horses can appear to be asymptomatic but will transmit the gene to 50% of their foals regardless of the other parent.

Genetics of roaring disorder in Thoroughbreds?

Q: Two years ago my father retired and now he is pursuing his lifelong passion to breed Thoroughbreds. His grandfather bred race horses in California about 100 years ago and he has always dreamed of restoring the family tradition. He purchased three older broodmares chosen for their special pedigrees and production records. He has studied bloodlines for years, from the standpoint of betting on races, but he has never before owned a horse or selected breeding pairs. A friend recently told him that he made a mistake in buying one of the mares because she has produced a roarer. This friend says roaring is a respiratory disease and a genetic trait. My sister and I are nurses and we are trying to help my father find out about this problem. Can you help us?

A: Roaring is a nervous system disorder that leads to paralysis of the left side of the larynx. The partial obstruction of air flow causes affected horses to make a whistling or roaring noise with physical exertion and compromises racing performance. This is classically a disease of Thoroughbreds but is not confined to that breed. Although commonly thought to be a genetic disease, breeding trials have not defined a mode of inheritance unequivocally. Cook (1978) reported roarers from matings of normal parents, thus suggesting a recessive trait, but two matings between roarer parents produced one normal and one affected foal, not consistent with a simple recessive gene model. Clearly more studies are needed to define the traits inheritance, including the possibility that it is a polygenic trait.

If you think your father's friend accurately reported about the mare's roarer foal, the best advice we can give you for your father's program is to try to avoid breeding that mare to a stallion that has offspring with the trait. This advice conservatively assumes that the trait is genetically controlled, but you will want to be on the lookout for any new research that provides definitive evidence about the trait's inheritance. Since recurrent breed disorders are generally associated with prominent bloodlines, finding a stallion may not be an easy task and, in addition, the information you seek may not be readily obtained. Unfortunately, without a defined mode of inheritance or a genetic test for trait carriers, it is very difficult to make informed breeding decisions that avoid the production of defective foals. Horse breeding can be a creative and worthwhile activity. Your father will need to decide whether he can economically and emotionally afford to breed horses considering all the genetic uncertainties.

Wobbler colt?

Q: I have a yearling gelding by an imported Dutch Warmblood stallion out of a 16-year-old Thoroughbred mare. I have used this mare very successfully in sport horse competition. Her yearling son is big, correct, handsome and just what I was looking for to replace his mother as my competition horse. He is turned out in pasture with other yearlings and for the last month he hasn't been moving right. He is clumsy and incoordinated. My veterinarian says he is ataxic due to a spinal cord injury of unknown origin. From what I can read, the "wobbler" problem could be caused by cervical vertebral fracture (trauma) or by vertebral malformation (possibly genetic). The stallion has lots of foals and the owner knows of no other wobblers. She has offered to rebreed my mare for free. Assuming the problem is genetic, what are my chances of getting another wobbler?

A: As you are aware, injury has not been ruled out as the source for your colt's problem and if a fracture can be demonstrated the genetic question is a moot point for you. However, the genetic issue occurs frequently. The literature on the genetics of wobbler syndrome is contradictory in several points. It is agreed by all studies that the problem is not inherited as a simple dominant and that wobbler × wobbler matings do not produce all wobbler offspring, thus ruling out a recessive hypothesis. Males are much more often affected than females, but an affected male bred to mares who had previously produced affected offspring did not produce wobbler foals, so the problem is not due to an X-linked gene. The thorough genetic analysis by Falco and colleagues (1976) failed to find evidence to support a genetic hypothesis for wobbler syndrome, at least for Thoroughbreds in Britain, and suggested that environmental factors should be closely examined as possible causative influences for the disease.

Prenatal testing as a tool for genetic selection in horses?

Q: I would like to get a tobiano foal from my tobiano Paint mare. From the standpoint of conformation I would like to breed her to my palomino QH stallion, but I realize my chances of getting a tobiano foal are only 50%. Once the mare is pregnant could I do any tests of her blood or maybe do amniocent-

esis to tell me if the foal is solid so I could abort the pregnancy and try again for a tobiano?

A: Prenatal testing is increasingly used in human medicine for pregnancies at risk for a deleterious genetic disease. Potentially the same kinds of procedures could be applied to horses, but practical considerations may make them unrealistic.

Termination of a pregnancy in horses must occur very early in gestation so that the mare will return to cycling within a few weeks. If the embryo develops beyond about 40 days, endometrial cup formation may prevent the mare from returning to estrus for several months so you may not get a foal for that season.

We are not aware of any blood tests of the *mare* in early pregnancy that would tell you about the genes of the embryo she is carrying. With amniocentesis you could obtain material from the offspring and use DNA technology to screen for genes whose DNA sequence information was available. At present, the only genes to which this could be applied are HYPP and sex, not tobiano, although the list is likely to be expanded in the near future. Another limitation would be finding a veterinarian skilled at amniocentesis in horses, but approach an equine reproduction specialist for a good place to start your inquiries.

If a DNA test for tobiano were available, embryo transplantation is another route that you might consider for your project. Several clinics nationwide provide such services for horses, although I am not aware that they are currently using genetic screening to select embryos. Using this route, you would recover an embryo from the mare soon after fertilization and test a few cells with the appropriate DNA technology. If the embryo was not of the desired genetic type, you could try again with another embryo from the mare's subsequent cycle. An acceptable embryo could be reintroduced into the donor or into any appropriate recipient mare to complete the pregnancy to term.

Carrier testing with random matings?

Q: I am very interested in breeding my American Saddlebred mare to a young stallion that has the qualities of bone and scope she needs. This horse only has seven foals on the ground. My mare has exceptional movement and attitude, but is possibly the carrier of an autosomal recessive lethal gene. I do not want to do this breeding if the stallion carries the same gene. How many normal foals from random matings are needed to prove that a stallion is not a carrier of this defective gene?

A: Random mating is not a very effective test of carrier status, but often it is the only information available. The answer depends on the gene frequency of the trait of concern. If 20% of the population carry a gene for an autosomal recessive undesirable trait, then about 60 unaffected foals from random matings are needed to provide statistical assurance at the 95% level that the stallion is not a carrier. If the trait is rare, say a carrier frequency of 2%, then 600 unaffected foals from random matings would be needed for 95% statistical assurance of non-carrier status of the sire.

Preventing NI problems?

Q: I own a pregnant Thoroughbred mare that I purchased two years ago. Her previous owners told me she had lost a foal to neonatal isoerythrolysis (NI). How can I have her tested to see if this foal is at risk? Can you also help me find a stallion that I can breed her to so I won't have to worry about NI at all?

A: A serum sample from the pregnant mare taken about three weeks before she is due to foal can be screened to see if a blood group incompatibility is detected that could lead to destruction of the foal's red blood cells (RBCs) and lead to death. If the test results are positive for anti-blood group activity, you should be prepared to attend the birth of the foal, making certain the foal does not suckle its dam and providing it with an alternative colostrum source. The foal can be put back to its dam's milk after 36–48 hours.

To prevent the problem in future foals, you can try to locate a stallion negative for the problem blood factor. For example, if the mare has anti-Aa antibodies, you could anticipate no NI problems in subsequent foals if the mare is bred to a stallion lacking the Aa factor. However, the Aa factor is very common; it may be difficult to find a stallion that lacks it and would suit in other ways. If the problem is Qa (or any other specificity), it will probably not be as difficult to find a blood group compatible stallion.

It may not be possible to find a stallion that is also suitable in pedigree, location, stud fee and conformation. In that case, you will need to manage the newborn foal as outlined above.

Infertility in a filly born cotwin to a colt?

Q: I have a yearling Quarter Horse filly that was twin to a colt that died at birth. My father tells me that in cattle, heifers born cotwin to bull calves are nearly always infertile. Is this also a possibility for horses?

A: This cattle condition is called "freemartinism." The biological mechanism that produces the freemartin is not known, but it is related to the masculinizing effect of shared blood circulation between twins of unlike sex that renders the heifer sterile. Evidence of shared circulation in horse twins has been reported from studies of fetal membranes after birth and confirmed by blood group chimerism and mixed XX/XY karyotypes from blood cultures of twin-born horses of unlike sex, but no cases of a masculinizing effect on the filly have been reported.

We have looked for a freemartin effect in horses using studbook data provided by the Arabian Horse Registry of America. We compared registrations for foals from twin-born mares in a female–female (FF) pair with foals from mares in a male–female (MF) pair. Among the 35 MF pairs occurring up to registration number 200,000 (foaling date up to about 1978), 24 mares had registered foals and 11 had none. Among the 19 FF twin pairs (38 mares), 30 had registered foals and eight had none. The difference between these groups is not statistically significant.

So the available evidence seems to be that the freemartin effect found in cattle is not of major concern for horse breeders. Infertility is not expected for a filly born cotwin to a colt.

Parentage testing, relatedness and pedigrees

Deriving genetic markers for a dead horse?

Q: My Morgan mare died before I had her blood typed. To register her 1994 foal I need to verify its parentage to both the dam and the sire. I have two full sisters to the 1994 foal. Can the type of the 1994 foal be compared with theirs to satisfy the parentage verification requirements?

A: Two offspring of an untested parent are unlikely to be sufficient to define its allelic pairs for the standard 15 system array of blood groups and proteins. To derive the genetic type of a dead horse at least 15–20 offspring are usually needed. Fewer offspring may be sufficient if their other parent or the parents of the dead horse can be tested.

If the dead mare was buried, it is possible that DNA testing from teeth or bone can be used to obtain a genetic profile for direct parentage validation. If the DNA route is chosen, the stallion and foal will need to be tested for these markers as well—a parentage analysis cannot be accomplished with blood group and protein markers for some animals in the case and DNA for others. Be sure to check with your registrar to see what kinds of genetic testing evidence is acceptable. He/she may help you access other genetic records that could help your derivation project.

Preservation of breeding lines and genetic diversity?

Q: My family has raised registered Arabian horses on our cattle ranch for about 50 years. My grandfather read extensively about the history and tradition of Arab horse breeding before selecting his horses from England and related horses from the USA. My uncle worked as an oil geologist in Saudi Arabia and brought back two mares from there for my father's breeding program. Now I am in line to direct this program. I took a degree in English history, without training in science or genetics. I do not want to change these horses. They suit our needs. Although we have never taken a horse to a show, a couple that we have sold have competed successfully both in the show ring and in performance. Most of our horses are in some degree related to each other and perhaps it is time to add new breeding stock. I have been reading Arabian horse magazines and have visited some "big-name" stables but I do not feel that the current show horse Arabian is what my grandfather had in mind when he set up our program. Am I right in thinking that I can preserve our stock as a genetically healthy group of Arabians without joining the mainstream?

A: Modern animal science principles encourage selection of breeding stock based on measured excellence (racing speed, show championships, milk production, egg production and so on). For your program such criteria may not be predictors of excellence. Since your horses suit your family's needs, you are certainly justified to stay with them. Your group is relatively little inbred at the moment and you may not need to add outside stock to maintain the program for many more years. If you did want to add horses one possibility may be to find horses related to your grandfather's early imports from England or your uncle's more recent desert imports.

We cannot know what uses will be made of horses in the twenty-first century, but twentieth century history has shown that narrow specialization can doom a breed. It is important to maintain diversity even in—or particularly in—the context of closed studbooks. Preservation programs are providing a strong voice for conserving breeding options. You are in a prime position to contribute to and benefit from such movements. If you choose this route for your breeding program, be prepared for skepticism by mainstream breeders, but be assured that the genetic basis is no less valid in the long term view than a program based on racetrack or arena winnings.

Half-siblings or random horses?

Q: I have two yearling horses, a filly and a colt, that I acquired from the US Department of Interior's Bureau of Land Management Adopt-A-Horse program for wild horses in Idaho. I understand these horses were captured at the same trap site. Can you do some genetic tests to tell me if they have the same sire? I am considering that I might breed the colt to the filly, but not if they are half-siblings.

A: If your filly and colt have sound conformation and meet your performance criteria, their potential relatedness would probably not be any more likely to produce foals with problems than would two horses from a breed where relationships have been recorded for generations. Genetic marker tests may be used to explore the possible relatedness of your pair, but the answer may not be in the form of a simple "yes" or "no." We cannot presume to know the results before the testing is accomplished, but let's review the kinds of data we may find and how we could interpret them.

Identical genetic markers in two horses could arise by direct descent from the same parent, by common descent from the same founders, or could be identical in phase (type) by chance. In a small population, such as feral horses in a herd management area, horses are likely to share markers and distant relationships to founder animals. A genetic profile of a random sample, say 50 animals, can provide an average value for identical markers in randomly chosen animals from this population.

Parent–offspring pairs and full-sib pairs are expected to share 50% of their genetic markers by direct inheritance. The observed value will be higher than 50% by the extent to which any two horses in the population share identical markers. In our studies with purebred horses, first generation relatives typically share 60–70% of genetic markers. Indeed, ostrich breeders use genetic marker testing to identify such pairs and avoid close relatedness.

Considering their ages, your pair could not be related as full siblings or as parent and offspring, so their expected levels of marker sharing put them in a more difficult position to evaluate. Half-siblings would have about 25% of their markers in common by direct descent. As with parent–offspring and full-sib pairs, the percentage of shared markers observed between half-siblings would most likely be greater than that expected solely on the basis of direct descent. If your pair had fewer than 25% identical markers it is possible that they could be related as half-sibs, *but highly unlikely.* What if their marker identity was 35%? If the population average was 15% or higher (which is highly likely), your pair would fall into a category for which the data could not determine their most probable relationship.

Genetic reconstruction of a highly regarded stallion?

Q: I hope your genetic marker testing can help me breed a genetic replica of a wonderful old Quarter Horse stallion that died two years ago. I have a daughter of this horse. I understand that she only got half of his chromosomes. Could your tests help me find a son for breeding with my mare that got the alternative chromosomes? I know I may have to repeat the mating several times and I know that my chances are not 100% to exactly match the old horse. The books say you need to use the young stallions to make genetic progress, but for my program the old-time genes are just fine and I can consistently produce just the kind of horse I need.

A: Unfortunately the Grand Genetic Game Plan is working against you in any scheme to *breed* a clone of the old stallion. As you know, the old stallion could only contribute half his genes to each foal. Your goal would be a double grandson to which its parents transmitted all the genes they received from the old stallion and none of the genes they received from their dams. Even if you could find a son of this old stallion that received the "other half" of his genes than your mare, it is probably impossible for you to raise enough foals from them to meet the statistical challenge of your proposal.

Another important point of discussion is to consider the number of genes in the genome. With our genetic markers we could easily identify 30 genes and may be another 20 more, but that is probably less than one thousandth the number of genes in any horse. Thus any genetic testing performed at this time unfortunately would not even be effective to help you identify those potential mates for the first step of your project.

Genetic planning is clearly possible at the level of selection for a few genes where the genetics is defined, such as breeding for certain coat colors. For conformation and performance traits where the genetics is much less defined and hundreds or thousands of genes are involved, the outcomes are much less predictable. Your skill as a breeder is determined by luck and by an intuitive understanding of the breeding consistencies of horses familiar to you. Science can help you plan matings for simple defined traits and predict the probabilities of the various outcomes, but the larger picture is still up to your vision.

Finding an identity for a rescued horse?

Q: I purchased a mare from the Society for the Prevention of Cruelty to Animals that had been rescued from a situation of poor care and near starvation. She is said to be a registered Morgan, but we do not have any specific details. She is bay with a star, about 10 years old. Could you look at her genetic markers and tell me who she is so I can get the papers for her?

A: You need to work with the American Morgan Horse Association to see what would be possible under their rules and regulations. If the AMHA considered the project to be appropriate, the Morgan database could be searched to look for a horse with genetic markers that match those of your horse. Of course age, sex, color and white markings would need to match as well. Probably the greatest obstacle for your project is that until very recently, only Morgan stallions had their genetic markers routinely recorded. Currently it is a requirement that all foals be parentage verified to both parents, so the vast majority of breeding stock has genetic markers recorded. If your new mare has not recently had a foal she may not have been tested previously and her identity could not be traced through this means of genetic matching.

Fingerprinting a stolen horse?

Q: My Quarter Horse colt was stolen a year ago and finally I have located him. Could you do a DNA fingerprint to compare him against his sire and prove that he is my horse? His dam was also stolen so I cannot get any blood samples from her, but I own the Thoroughbred sire and can send you hair or blood or whatever samples you need.

A: Fingerprinting has been a useful technique for forensic laboratories attempting to identify crime scene samples, but is not as useful for looking at potential genetic relationships between a pair of individuals. You probably used the term fingerprinting because you hear it used in the popular press and assume that it is a general term for genetic marker testing. The most powerful tools in horses for looking at the relationship between an offspring and a putative parent are genetic systems whose alleles have been clearly defined, such as blood groups, protein polymorphisms and microsatellites. Allelic definition allows us to apply the power of the Mendelian law of segregation to identify incorrect paternity or maternity. Using a battery of such tests the probability of exclusion of a randomly assigned (false) parent in Thoroughbreds exceeds 99.5%. If your pair is genetically compatible in all systems and you can show that the found horse is identical to the stolen one in age, color, markings and sex these data can be presented as compelling evidence that the recovered horse is the same as the one stolen from you a year ago.

Chapter 20
Where are we going from here?

Horse genetics is a superb candidate to benefit from the molecular technology revolution. Traditional methods for genetics research that rely on collecting breeding data are slow to yield results in horses because of small "litter" size and a relatively long gestation period. Often the genetic experiments never get started because the expensive maintenance for the project horses cannot be justified by the potential information gain. For some questions in horse genetics, molecular techniques may allow the use of basic information from other organisms, particularly human beings and mice. The initial research phases can be conducted from fresh, frozen or immortalized (cultured) horse tissue samples. The breeding trial stages might then ensue by using horses from cooperative breeders, without needing to include prospective breeding trials as part of the research budget.

The following areas of horse genetics are particularly likely to see dramatic changes:

- Parentage testing will increasingly incorporate techniques that assay specific regions of DNA sequences known to be genetically variable between individuals. Initially the phenotyping will use microsatellites. Recognizing the potential worldwide movement of horses and the useful model of conventional blood typing that allows reciprocity of test results between laboratories, active international dialog concerning standards and controls will assure that DNA typing results can be used in this context. Hair may replace blood as the tissue of choice for parentage testing. Basic research in horse mitochondrial sequences will allow specific recognition of false maternal assignment.
- Genetic mapping of the horse genome will proceed rapidly. The value of this progress may not be immediately obvious to breeders, but will allow the eventual assignment of specific DNA sequences to conspicuous phenotypes of horses (e.g. those of colors and diseases). Two kinds of genetic information will initially be used for mapping. Most of the markers will be microsatellites (generally genes of unknown function) specific to the horse. The second type of markers will be protein coding genes identified from other mammals as anchor loci for defining gene order and relationship to

conserved regions of the genetic maps of other species. These studies will be a worldwide endeavor, allowing maximum gain from the efforts involved. Genes will be assigned to syntenic groups using somatic cell hybrid panels. Linkage distances and gene orders will be defined from family studies and gene assignment to chromosomes will follow from FISH studies of selected genes within recognized linkage groups.

- Diagnostic tests for the potential presence of recessive genes will follow from the identification of DNA sequences. These tests will be particularly important for making breeding decisions using horses known to be carriers of genetic diseases but otherwise of excellent breed type and performance ability. Gene diagnostics will provide welcome information for color breeders.
- Identifying the relationships between breeds and related species will proceed apace, using a comprehensive panel of markers, including coding and non-coding genes of both nuclear and mitochondrial DNA.
- The interaction of nutrition and genetics will be slowly but steadily better defined as specific genes controlling traits are identified and their biochemical actions are understood. Appropriate nutritional strategies for management of genetic traits sensitive to dietary schemes can then be outlined.
- Performance traits specific to the horse and likely to be polygenic will be better understood once a comprehensive map of the horse genome is available. The map will be a tool for locating the important genes, and their subsequent identification may follow from comparative studies of maps of the horse with those of other organisms.

The explosion in genetic information specifically about horses means that breeders will want to have a firm understanding of genetic principles. They will want to be able to recognize research applications that are of value to their particular program and evaluate advertising hyperbole that will hope to sell them the technology. They will comprehend the results obtained from the newly developed tests and how to apply them to their goals of raising good horses. They will be able to evaluate the claims of other breeders about the superior genetics of their stock created with the new genetics and will know the limits of the potential gains. Above all, breeders with a firm understanding of genetics will understand that it takes a long time to effect genetic changes in a breeding program and will invoke the necessary patience. Good luck and have fun!

REFERENCES

Adalsteinsson, S., 1974. Inheritance of the palomino color in Icelandic Horses. Journal of Heredity 65:15–20.

Allen, W.R. and Pashen, R.L., 1984. Production of monozygotic (identical) horse twins by embryo micromanipulation. Journal of Reproduction and Fertility 71:607–613.

Andersson, L. and Sandberg, K., 1982. A linkage group composed of three coat color genes and three serum protein loci in horses. Journal of Heredity 73:91–94.

Andersson, L., Juneja, R.K. and Sandberg, K., 1983a. Genetic linkage between the loci for phosphohexose isomerase (*PHI*) and a serum protein (*Xk*) in horses. Animal Blood Groups and Biochemical Genetics 14:45–50.

Andersson, L., Sandberg, K., Adalsteinson, S. and Gunnarsson, E., 1983b. Linkage of the equine serum esterase (*Es*) and mitochondrial glutamate oxaloacetate transaminase (GOT_m) loci. Journal of Heredity 74:361–364.

Angrist, M., Kauffman, E., Slaugenhaupt, S.A., Matise, T.C., Puffenberger, E.G., Washington, S.S., Lipson, A., Cass, D.T., Reyna, T., Weeks, D.E., Sieber, W. and Chakravarti, A., 1993. A gene for Hirschsprung disease (megacolon) in the pericentromeric region of human chromosome 10. Nature Genetics 4:351–356.

Ansari, H.A., Hediger, R., Fries, R. and Stranzinger, G., 1988. Chromosomal localization of the major histocompatibility complex of the horse (*ELA*) by *in situ* hybridization. Immunogenetics 28:362–364.

Archer, R.K., 1961. True haemophilia (haemophilia A) in a Thoroughbred foal. Veterinary Record 73:338–340.

Bailey, E., 1984. Usefulness of lymphocyte typing to exclude incorrectly assigned paternity in horses. American Journal of Veterinary Research 45:1976–1978.

Bailey, E., Albright, D.G. and Henney, P.J., 1988. Equine neonatal isoerythrolysis: evidence for prevention by maternal antibodies to the Ca blood group antigen. American Journal of Veterinary Research 49:1218–1222.

Bailey, E., Graves, K.T., Cothran, E.G., Reid, R., Lear, T.L. and Ennis, R.B., 1995. Synteny-mapping horse microsatellite markers using a heterohybridoma panel. Animal Genetics 26:177–180.

Bailey, E., Stormont, C., Suzuki, Y. and Trommershausen-Smith, A., 1979. Linkage of loci controlling alloantigens on red blood cells and lymphocytes in the horse. Science 204:1317–1319.

Baldwin, C.T., Hoth, C.F., Amos, J.A., da-Silva, E.O. and Milunsky, A., 1992. An exonic mutation in the *HuP2* paired domain gene causes Waardenburg's syndrome. Nature 355:637–638.

Basrur, P.K., Kanagawa, H. and Gilman, J.P.W., 1970. Further studies on the cell populations of an intersex horse. Journal of Comparative Medicine 34:294–298.

Belyaev, D.K., 1979. Destabilizing selection as a factor in domestication. Journal of Heredity 70:301–308.

Bengtsson, S. and Sandberg, K., 1973. A method for simultaneous electrophoresis of four horse red cell enzyme systems. Animal Blood Groups and Biochemical Genetics 4:83–87.

Benirschke, K. and Ryder, O.A., 1985. Genetic aspects of equids with particular reference to their hybrids. Equine Veterinary Journal, Suppl. 3:1–10.

Bernoco, D. and Byrns, G., 1991. DNA fingerprint variation in horses. Animal Biotechnology 2:145–160.

Bernoco, D., Byrns, G., Bailey, E. and Lew, A.M., 1987a. Evidence of a second polymorphic ELA class I (ELA-B) locus and gene order for three loci of the equine major histocompatibility complex. Animal Genetics 18:1103–1118.

Bernoco, D., Antczak, D.F., Bailey, E., Bell, K., Bull, R.W., Byrns, G., Guerin, G., Lazary, S., McClure, J., Templeton, J. and Varewyck, H., 1987b. Joint report of the Fourth International Workshop on Lymphocyte Alloantigens of the Horse, Lexington, Kentucky, 12–22 October 1985. Animal Genetics 18:81–94.

Björck, G., Everz, K.E., Hansen, H.-J. and Henricson, B., 1973. Congenital cerebellar ataxia in the Gotland pony breed. Zentralblatt für Veterinarmedizin. Reihe A (Berlin) 20:341–354.

Blakeslee, L.H., Hudson, R.S. and Hunt, H.R., 1943. Curly coat of horses. Journal of Heredity 34:115–118.

Blue, M.G., 1981. A cytogenetical study of prenatal loss in the mare. Theriogenology 15:295–309.

Bowling, A.T., 1985. The use and efficacy of horse blood typing tests. Equine Veterinary Science 5:195–199.

Bowling, A.T., 1986. Genetic linkage between loci for a red cell alloantigen (*U*) and serum protease inhibitor (*PI*) in the horse. Animal Genetics 17:217–223.

Bowling, A.T., 1987. Equine linkage group II: phase conservation of *To* with Al^B and Gc^S. Journal of Heredity 78:248–250.

Bowling, A.T., 1992. Genetics of the horse. *In* Evans, J.W. (Editor), Horse Breeding and Management. Elsevier, New York, pp. 207–236.

Bowling, A.T., 1994a. Dominant inheritance of overo spotting in Paint horses. Journal of Heredity 85:222–224.

Bowling, A.T., 1994b. Population genetics of Great Basin feral horses. Animal Genetics 25, Suppl 1:67–74.

Bowling, A.T. and Clark, R.S., 1985. Blood group and protein polymorphism gene frequencies for seven breeds of horses in the United States. Animal Blood Groups and Biochemical Genetics 16:93–108.

Bowling, A.T. and Millon, L., 1988. Centric fission in the karyotype of a mother–daughter pair of donkeys (*Equus asinus*). Cytogenetics and Cell Genetics 47:152–154.

Bowling, A.T. and Millon, L., 1990. Two autosomal trisomies in the horse: 64,XX,–26+t(26q26q) and 65,XX,+30. Genome 33:679–682.

Bowling, A.T. and Ryder, O.A., 1987. Genetic studies of blood markers in Przewalski's horses. Journal of Heredity 78:75–80.

Bowling, A.T. and Williams, M.J., 1991. Expansion of the D system of horse red cell alloantigens. Animal Genetics 22:361–367.

Bowling, A.T., Millon, L. and Hughes, J.P., 1987. An update of chromosomal abnormalities in mares. Journal of Reproduction and Fertility, Suppl. 35:149–155.

Bowling, A.T., Scott, A.M., Flint, J. and Clegg, J.B., 1988. Novel alpha haemoglobin haplotypes in horses. Animal Genetics 19:87–101.

Bowling, A.T., Gordon, L., Penedo, M.C.T., Wictum, E. and Beebout, J., 1990. A single gel for determining genetic variants of equine erythrocyte carbonic anhydrase (*CA*) and catalase (*Cat*). Animal Genetics 21:191–197.

Braend, M., 1970. Genetics of horse acidic prealbumins. Genetics 65:495–503.

Breen, M., Downs, P., Irvin, Z. and Bell, K., 1994. Six equine dinucleotide repeats: microsatellites MPZ002, 3, 4, 5, 6 and 7. Animal Genetics 25:124.

Brilliant, M.H., Gondo, Y. and Eicher E.M., 1991. Direct molecular identification of the mouse pink-eyed unstable mutation by genome scanning. Science 252:566–569.

Buckland, R.A., Fletcher, J.M. and Chandley, A.C., 1976. Characterization of the domestic horse (*Equus caballus*) karyotype using G- and C-banding techniques. Experientia 32:1146–1149.

Castle, W.E. and Smith, F.H., 1953. Silver dapple, a unique color variety among Shetland ponies. Journal of Heredity 44:139–145.

Chabot, B., Stephenson, D.A., Chapman, V.M., Besmer, P. and Bernstein, A., 1988. The proto-oncogene c-kit encoding a transmembrane tyrosine kinase receptor maps to the mouse *W* locus. Nature 335:88–89.

Chandley, A.C., Fletcher, J., Rossdale, P.D., Peace, C.K., Ricketts, S.W., McEnery, R.J., Thorne, J.P., Short, R.V. and Allen, W.R., 1975. Chromosome abnormalities as a cause of infertility in mares. Journal of Reproduction and Fertility, Suppl. 23:377–383.

Chowdhary, B.P., Harbitz, I., Davies, W. and Gustavsson, I., 1992. Localization of the calcium release channel gene in cattle and horse by *in situ* hybridization: evidence of a conserved synteny with glucose phosphate isomerase. Animal Genetics 23:43–50.

Clutton-Brock, J., 1992. Horse Power. Harvard University Press. Cambridge.

Constant, S.B., Larsen, R.E., Asbury, A.C., Buoen, L.C. and Mayo, M., 1994. XX male syndrome in a cryptorchid stallion. Journal of the Veterinary Medical Association 205:83–85.

Cook, W.R., 1978. Hereditary diseases in the horse. *Reference cited in* Equine Genetics and Selection Procedures, L.W. Chalkley and W.R. Cook (Editors), 1978, Equine Research Publications, Dallas.

Copeland, N.G., Jenkins, N. and Lee, B.K., 1983a. Association of the lethal yellow (A^y) coat color mutation with an exotropic murine leukemia virus genome. Proceedings of the National Academy of Sciences USA 80:247–249.

Copeland, N.G., Hutchison, K.W. and Jenkins, N.A., 1983b. Excision of the DBA ecotropic provirus in dilute coat-color revertants of mice occurs by homologous recombination involving the viral LTRs. Cell 33:379–387.

Cothran, E.G., MacCluer, J.W., Weitkamp, L.R. and Bailey, E., 1987. Genetic differentiation associated with gait within American Standardbred horses. Animal Genetics 18:285–296.

Cothran, E.G., MacCluer, J.W., Weitkamp, L.R., Pfenning, D.W. and Boyce, A.J., 1984. Inbreeding and reproductive performance in Standardbred horses. Journal of Heredity 75:220–224.

Cox, J.H., 1985. An episodic weakness in four horses associated with intermittent serum hyperkalemia and the similarity of the disease to hyperkalemic periodic paralysis in man. *In* Proceedings, American Association of Equine Practitioners, 1985, pp. 383–391.

Cunningham, E.P., 1975. Genetic studies in horse populations. *In* Proceedings, International Symposium on Genetics and Horse-Breeding. Royal Dublin Society, September, 1975, pp. 2–6.

DeBowes, R.M., Leipold, H.W. and Turner-Beatty, M., 1987. Cerebellar abiotrophy. Veterinary Clinics of North America: Equine Practice 3:345–352.

Dreux, P., 1966. Introduction statistique à la génétique des marques blanches limitées chez le cheval domestique. Annales de Génétiques 9:66–72.

Dunn, H.O., Smiley, D., Duncan, J.R. and McEntee, K., 1981. Two equine true hermaphrodites with 64,XX/64,XY and 63,X/64,XY chimerism. Cornell Veterinarian 71:123–135.

Dyke, T.M., Laing, E.A. and Hutchins, D.R., 1990. Megacolon in two related Clydesdale foals. Australian Veterinary Journal 67:463–464.

Ellegren, H., Johansson, M., Sandberg, K. and Andersson, L., 1992. Cloning of highly polymorphic microsatellites in the horse. Animal Genetics 23:133–142.

Ellersieck, M.R., Lock, W.E., Vogt, D.W. and Aipperspach, R., 1985. Genetic evaluation of cutting scores in horses. Equine Veterinary Science 5:287–289.

Eriksson, K., 1955. Hereditary aniridia with secondary cataract in horses. Nordisk Veterinarmedicin 7:773–793.

Ewen, K.R. and Matthews, M.E., 1994. Equine dinucleotide repeat polymorphism at the VIAS-H7 locus. Animal Genetics 25:292.

Falco, M.J., Whitwell, K. and Palmer, A.C., 1976. An investigation into the genetics of "wobbler" disease in Thoroughbred horses in Britain. Equine Veterinary Journal 8:165–169.

Gaffney, B. and Cunningham, E.P., 1988. Estimation of genetic trend in racing performance of Thoroughbred horses. Nature 332:722–724.

Gardner, E.J., Shupe, J.L., Leone, N.C. and Olson, A.E., 1975. Hereditary multiple exostoses. Journal of Heredity 66:318–322.

Gaughan, E.M. and DeBowes, R.M., 1993. Congenital diseases of the equine head. Veterinary Clinics of North America: Equine Practice 9: 93–110.

Geissler, E.N., Ryan, M.A. and Housman, D.E., 1988. The dominant-white spotting (*W*) locus of the mouse encodes the *c-kit* proto-oncogene. Cell 55:185–192.

Georges, M., Lequarré, A.-S., Castelli, M., Hanset, R. and Vassart, G., 1988. DNA fingerprinting in domestic animals using four different minisatellite probes. Cytogenetics and Cell Genetics 47:127–131.

Gerlach, J.A., Bowling, A.T., Bowling, M. and Bull, R.W., 1994. Assignment of maternal lineage using mitochondrial nucleic acid sequence in horses. Animal Genetics 25, Suppl. 2: 31.

Gerneke, W.H. and Coubrough, R.I., 1970. Intersexuality in the horse. Onderstepoort Journal of Veterinary Research 37:211–216.

Geurts, R., 1977. Hair Colour in the Horse. J.A. Allen, London.

Gill, P., Ivanov, P., Kimpton, C., Piercy, R., Benson, N., Tully, G., Evett, I., Hagelberg, E. and Sullivan, K., 1994. Identification of the remains of the Romanov family by DNA analysis. Nature Genetics 6:130–135.

Greene, M.C. (ed.), 1981. Genetic Variants and Strains of the Laboratory Mouse. Gustav Fischer Verlag, New York.

Gremmel, F., 1939. Coat color in horses. Journal of Heredity 30:437–445.

Grøndahl, A.M. and Dolvik, N.I., 1993. Heritibility estimations of osteochondrosis in the tibiotarsal joint and of bony fragments in the palmar/plantar portion of the metacarpo- and metatarsophalangeal joints of horses. Journal of the American Veterinary Medical Association 203:101–104.

Groves, C. P., 1974. Horses, Asses and Zebras in the Wild. Ralph Curtis Books, Hollywood, Florida.

Gu, F., Harbitz, I., Chowdhary, B.P., Chowdhary, R. and Gustavsson, I., 1992. Localization of the 6-phosphogluconate dehydrogenase (PGD) gene in horses by *in situ* hybridization. Hereditas 117:93–95.

Guérin, G., Bertaud, M. and Amigues, Y., 1994. Characterization of seven new horse microsatellites: HMS1, HMS2, HMS3, HMS5, HMS6, HMS7 and HMS8. Animal Genetics 25:62.

Gundlach, A.L., Kortz, G., Burazin, T., Madigan, J. and Higgins, R.J., 1993. Deficit of inhibitory glycine receptors in spinal cord from Peruvian Pasos evidence for an equine form of inherited myoclonus. Brain Research 628:263–270.

Hall, B.R., 1995. Atavisms and atavistic mutations. Nature Genetics 10:126–127.

Harbitz, I., Chowdhary, B.P., Saether, H., Hauge, J.G. and Gustavsson, I., 1990. A porcine genomic glucosephosphate isomerase probe detects a multiallelic restriction fragment length polymorphism assigned to chromosome 10pter in horse. Hereditas 112:151–156.

Hardy, M.H., Fisher, K.R.S., Vrablic, O.E., Yager, J.A., Nimmo-Wilkie, J.S., Parker, W. and Keeley, F.W., 1988. An inherited connective tissue disease in the horse. Laboratory Investigation 59:253–262.

Hatami-Monazah, H. and Pandit, R.V., 1979. A cytogenetic study of the Caspian pony. Journal of Reproduction and Fertility 57:331–333.

Haynes, S.E. and Reisner, A.H., 1982. Cytogenetic and DNA analyses of equine abortion. Cytogenetics and Cell Genetics 34:204–214.

Hedrick, P.W., 1983. Genetics of Populations. Van Nostrand Reinhold, New York.

Henninger, R.W., 1988. Hemophilia A in two related Quarter Horse colts. Journal of the American Veterinary Medical Association 193:91–94.

Hermans, W.A., 1970. A hereditary anomaly in Shetland ponies. Netherlands Journal of Veterinary Science 3:55–63.

Hermans, W.A., Kersjes, A.W., van der Mey, G.J.W. and Dik, K.J., 1987. Investigation into the heredity of congenital lateral patellar (sub)luxation in the Shetland pony. Veterinary Quarterly 9:1–8.

Hintz, H.F. and Van Vleck, L.D., 1979. Lethal dominant roan in horses. Journal of Heredity 70:145–146.

Hintz, R.L., 1980. Genetics of performance in the horse. Journal of Animal Science 51:582–594.

Hughes, J.P., Benirschke, K., Kennedy, P.C. and Trommershausen-Smith, A., 1975. Gonadal dysgenesis in the mare. Journal of Reproduction and Fertility, Suppl. 23:385–390.

Hultgren, B.D., 1982. Ileocolonic aganglionosis in white progeny of overo spotted horses. Journal of the American Veterinary Medical Association 180:289–292.

Hurst, C.C., 1906. On the inheritance of coat colour in horses. Proceedings of the Royal Society, Series B 77:388–394.

Hutchins, D.R., Lepherd, E.E. and Crook, I.G., 1967. A case of equine haemophilia. Australian Veterinary Journal 43:83–87.

Ishida, N., Hasegawa, T., Takeda, K., Sakagami, M., Onishi, A., Inumaru, S., Komatsu, M. and Mukoyama, H., 1994. Polymorphic sequence in the D-loop region of equine mitochondrial DNA. Animal Genetics 25:215–221.

Jamieson, A., 1965. The genetics of transferrins in cattle. Heredity 20:419–441.

Jamison, J.M., Baird, J.D., Smith-Maxie, L.L. and Hulland, T.J., 1987. A congenital form of myotonia with dystrophic changes in a Quarterhorse. Equine Veterinary Journal 19:353–358.

Johnson, G.C., Kohn, C.W., Johnson, C.W., Garry, F., Scott, D. and Martin, S., 1988. Ultrastructure of junctional epidermolysis bullosa in Belgian foals. Comparative Pathology 98:329–336.

Juneja, R.K., Gahne, B. and Sandberg, K., 1978. Genetic polymorphism of the vitamin D binding protein and another post-albumin protein in horse serum. Animal Blood Groups and Biochemical Genetics 9:29–36.

Kay, P.H., Dawkins, R.L., Bowling, A.T. and Bernoco, D., 1987a. Heterogeneity and linkage of equine *C4* and *steroid 21-hydroxylase* genes. Journal of Immunogenetics 14:247–253.

Kay, P.H., Dawkins, R.L., Bowling, A.T. and Bernoco, D., 1987b. Polymorphism of the acetylcholine receptor in the horse. Veterinary Record 120:363–365.

Kelly, E.P., Stormont, C. and Suzuki, Y., 1971. Catalase polymorphism in the red cells of horses. Animal Blood Groups and Biochemical Genetics 2:135–143.

Kent, M.G., Shoffner, R.N., Buoen, L. and Weber, A.F., 1986. XY sex-reversal syndrome in the domestic horse. Cytogenetics and Cell Genetics 42:8–18.

Kent-First, M.G., Buoen, L., Zhang, T.Q., Meisner, L.F. and Nolten, W.E., 1995. Comparative deletion analysis of Y chromosome genes which determine gonadal development and function in XX male humans, horses and dogs. *In* Proceedings, 9th North American Colloquium on Domestic Animal Cytogenetics and Gene Mapping, College Station, p. 25.

Kieffer, N.M., Burns, S.J. and Judge, N.G., 1976. Male pseudo-hermaphroditism of the testicular feminizing type in a horse. Equine Veterinary Journal 8:38–41.

Klemetsdal, G., 1990. Breeding for performance traits in horses—A review. *In* Proceedings, 4th World Congress on Genetics Applied to Livestock Production 16:184–193.

Klemola, V., 1933. The "pied" and "splashed white" patterns in horses and ponies. Journal of Heredity 24:65–69.

Klunder, L.R., McFeely, R.A., Beech, J., McClune, W. and Bilinski, W.F., 1989. Autosomal trisomy in a Standardbred colt. Equine Veterinary Journal 21:69–70.

Knottenbelt, D. and Pascoe, R., 1994. Color Atlas of Diseases and Disorders of the Horse. Wolfe (Mosby-Year Book Europe Limited), London.

Koch, P. and Fischer, H., 1951. Die Oldenburger Fohlenataxie als Erbkrankheit. Tierarztliche Umschau 6:158–159.

Kopp, E., Mayr, B., Czaker, R. and Schleger, W., 1981. Nucleolus organizer regions in the domestic horse. Journal of Heredity 72:357–358.

Kopp, E., Mayr, B., Kalat, M. and Schleger, W., 1988. Polymorphisms of NORs and heterochromatin in the horse and donkey. Journal of Heredity 79:332–337.

Kubien, E.M., Pozor, M. and Tischner, M., 1993. Clinical, cytogenetic and endocrine evaluation of a horse with a 65,XXY karyotype. Equine Veterinary Journal 24:333–335.

Lane, P.W. and Liu, H.M., 1984. Association of megacolon with a new dominant spotting gene (*Dom*) in the mouse. Journal of Heredity 75:435–439.

Langemeier, J.L., Bailey, E. and Henney, P.J., 1993. Linkage studies between *Tcp-1*, *Tcp-10*, and *MHC-Eqca-A* loci in the horse. Immunogenetics 38:359–362.

Langlois, B., 1980. Heritability of racing ability in Thoroughbreds—A review. Livestock Production Science 7:591–605.

Lazary, S., Gerber, H., Glatt, P.A. and Straub, R., 1985. Equine leucocyte antigens in sarcoid-affected horses. Equine Veterinary Journal 17:283–286.

Legault, C., 1976. Can we select systematically for jumping ability? *In* Proceedings, International Symposium on Genetics and Horse-Breeding (Dublin, 1975), pp. 71–77.

Leipold, H.W., Brandt, G.W., Guffy, M. and Blauch, B., 1974. Congenital atlanto-occipital fusion in a foal. Veterinary Medicine: Small Animal Clinics 69:1312–1316.

Lerner, D.J. and McCracken, M.D., 1978. Hyperelastosis cutis in 2 horses. Journal of Equine Medicine and Surgery 2:350–352.

Lew, A.M., Bailey, E., Valas, R.B. and Coligan, J., 1986. The gene encoding the equine soluble class I molecule is linked to the horse MHC. Immunogenetics 24:128–130.

Long, S.E., 1994. A tandem translocation in the horse. *In* Proceedings, 11th European Colloquium on Cytogenetics of Domestic Animals, Copenhapen, pp. 92–94.

Lunn, D.P., Cuddon, P.A., Shaftoe, S. and Archer, R.M., 1993. Familial occurrence of narcolepsy in Miniature horses. Equine Veterinary Journal 25:483–487.

Lush, J.L., 1945. Animal Breeding Plans. Iowa State University Press, Ames.

Lyonnet, S., Bolino, A., Pelet, A., Abel, L., Nihoul-Fékété, C., Briard, M.L., Mok-Siu, V., Kaariainen, H., Martucciello, G., Lerone, M., Puliti, A., Luo, Yin, Weissenbach, J., Devoto, M., Munnich., A. and Romero, G., 1993. A gene for Hirschsprung disease maps to the proximal long arm of chromosome 10. Nature Genetics 4:346–350.

MacCluer, J.W., Boyce, A.T., Dyke, B., Weitkamp, L.R., Pfennig, D.W. and Parsons, C.J., 1983. Inbreeding and pedigree structure in Standardbred horses. Journal of Heredity 74:394–399.

Mäkinen, A., Chowdhary, B.P., Mahdy, E., Andersson, L. and Gustavsson, I., 1989. Localization of the equine major histocompatibility complex (ELA) to chromosome 20 by *in situ* hybridization. Hereditas 110:93–96.

Marklund, S., Ellegren, H., Eriksson, S., Sandberg, K. and Andersson, L., 1994. Parentage testing and linkage analysis in the horse using a set of highly polymorphic microsatellites. Animal Genetics 25:19–23.

Mayhew, I.G., Brown, C.M., Stowe, H.D., Trapp, A.L., Derksen, F.J. and Clement, S.F., 1987. Equine degenerative myeloencephalopathy: a vitamin E deficiency that may be familial. Journal of Veterinary Internal Medicine 1:45–50.

Mayhew, I.G., Watson, A.S. and Heissan, J.A., 1978. Congenital occipito-atlanto-axial malformations in the horse. Equine Veterinary Journal 10:103–113.

McCabe, L., Griffin, L.D., Kinzer, A., Chandler, M., Beckwith, J.B. and McCabe, R.B., 1990. Overo lethal white foal syndrome: equine model of aganglionic megacolon (Hirschsprung Disease). American Journal of Medical Genetics 36:336–340.

McGuire, T.C., Banks, K.L., Evans, D.R. and Poppie, M.J., 1976. Agammaglobulinemia in a horse with evidence of functional T lymphocytes. Journal of Veterinary Research 37:41–46.

McGuire, T.C., Poppie, M.J. and Banks, K.L., 1974. Combined (B- and T-lymphocyte) immunodeficiency: a fatal genetic disease in Arabian foals. Journal of the American Veterinary Medical Association 164:70–76.

McIlwraith, C.W., Owen, R. and Basrur, P.K., 1976. An equine cryptorchid with testicular and ovarian tissues. Equine Veterinary Journal 8:156–160.

McKusick, V.A., 1990. Mendelian Inheritance in Man. 9th Edition. Johns Hopkins University Press, Baltimore.

Millon, L.V., Bowling, A.T. and Bickel, L.A., 1993. Fluorescence *in situ* hybridization of C3 and 18S rDNA to horse chromosomes. Proceedings of the 8th North American Colloquium on Domestic Animal Cytogenetics and Gene Mapping, Guelph, p. 163.

Nicholas, F.W., 1987. Veterinary Genetics. Clarendon Press, Oxford.

Nikiforov, T.T., Rendle, R.B., Goelet, P., Rogers, Y-H., Kotewicz, M.L., Anderson, S., Trainor, G.L. and Knapp, M.R., 1994. Genetic Bit Analysis: a solid phase method for typing single nucleotide polymorphisms. Nucleic Acids Research 22:4167–4175.

Noda, H. and Watanabe, Y., 1975. Relationship between blood groups and hemolytic disease of newborn foals. Japanese Journal of Zootechnical Science 46:180–184.

Oakenfull, E.A., Buckle, V.J. and Clegg, J.B., 1993. Localization of the horse (*Equus caballus*) α-globin gene complex to chromosome 13 by fluorescence *in situ* hybridization. Cytogenetics and Cell Genetics 62:136–138.

O'Brien, S.J., Womack, J.E., Lyons, L.A., Moore, K.J., Jenkins, N.A and Copeland, N.G, 1993. Anchored reference loci for comparative genome mapping in mammals. Nature Genetics 3:103–112.

Odriozola, M., 1951. A los Colores del Caballo. Madrid.

Palmer, A.C., Blakemore, W.F., Cook, W.R., Platt, H. and Whitwell, K.E., 1973. Cerebellar hypoplasia and degeneration in the young Arab horse: clinical and neuropathological features. Veterinary Record 93:62–66.

Pirchner, F., 1983. Population Genetics in Animal Breeding. 2nd Edition. Plenum Press, New York.

Pollitt, C. and Bell, K., 1980. Protease inhibitor system in horses: classification and detection of a new allele. Animal Blood Groups and Biochemical Genetics 11:235–244.

Poppie, M.J. and McGuire, T.C., 1977. Combined immunodeficiency in foals of Arabian breeding: evaluation of mode of inheritance and estimation of prevalence of affected foals and carrier mares and stallions. Journal of the American Veterinary Medical Association 170:31–33.

Power, M.M., 1987. Equine half sibs with an unbalanced X;15 translocation or trisomy 28. Cytogenetics and Cell Genetics 45:163–168.

Power, M.M., 1988. Y chromosome length variation and its significance in the horse. Journal of Heredity 79:311–313.

Power, M.M., 1991. The first description of a balanced reciprocal translocation t(1q;3q) and its clinical effects in a mare. Equine Veterinary Journal 23:146–149.

Pulos, W.L. and Hutt, F.B., 1969. Lethal dominant white in horses. Journal of Heredity 60:59–63.

Rando, A., Di Gregorio, P. and Masina, P., 1986. Polymorphic restriction sites in the horse β-globin gene cluster. Animal Genetics 17:245–253.

Rendel, J. and Gahne, B., 1961. Parentage tests in cattle using erythrocyte antigens and serum transferrins. Animal Production 3:307–314.

Richer, C.L., Power, M.M., Klunder, L.R., McFeely, R.A. and Kent, M.G., 1990. Standard karyotype of the domestic horse (*Equus caballus*). Hereditas 112:289–293.

Robbins, L.S., Nadeau, J.H., Johnson, K.R., Kelly, M.A., Roselli-Rehfuss, L., Baack, E., Mountjoy, K.G. and Cone, R.G., 1993. Pigmentation phenotypes of variant extension locus alleles result from point mutations that alter MSH receptor function. Cell 72:827–834.

Romagnano, A. and Richer, C.-L., 1984. R-banding of horse chromosomes. Journal of Heredity 75:269–272.

Rong, R., Chandley, A.C., Song, J., McBeath, S., Tan, P.P., Bai, Q. and Speed, R.M., 1988. A fertile mule and hinny in China. Cytogenetics and Cell Genetics 47:134–139.

Rossdale, R.D., 1972. Modern concepts of neonatal disease in foals. Equine Veterinary Journal 4:117–128.

Rudolph, J.A., Spier, S.J., Byrns, G., Rojas, C.V., Bernoco, D. and Hoffman, E.P., 1992. Periodic paralysis in Quarter Horses: a sodium channel mutation disseminated by selective breeding. Nature Genetics 2:144–147.

Ryder, O.A. and Chemnick, L.G., 1990. Chromosomal and molecular evolution in Asiatic wild asses. Genetica 83:67–72.

Ryder, O.A., Chemnick, L.G., Bowling, A.T. and Benirschke, K., 1985. Male mule foal qualifies as the offspring of a female mule and jack donkey. Journal of Heredity 76:379–381.

Ryder, O.A., Epel, N.C. and Benirschke, K., 1978. Chromosome banding studies of the Equidae. Cytogenetics and Cell Genetics 20:323–350.

Sakagami, M., Tozaki, T., Mashima, S., Hirota, K. and Mukoyama, H., 1995. Equine parentage testing by microsatellite locus at chromosome 1q2.1. Animal Genetics 26:123–124.

Sandberg, K., 1968. Genetic polymorphism in carbonic anhydrase from horse erythrocytes. Hereditas 60:411.

Sandberg, K., 1973. The D blood group system of the horse. Animal Blood Groups and Biochemical Genetics 4:193–205.

Sandberg, K., 1974a. Linkage between the K blood group locus and the 6-PGD locus in horses. Animal Blood Groups and Biochemical Genetics 5:137–141.

Sandberg, K., 1974b. Genetically controlled variants of NADH diaphorase from horse red cells. Animal Blood Groups and Biochemical Genetics 5, suppl. 1:23–24.

Sandberg, K. and Juneja, R.K., 1978. Close linkage between the albumin and *Gc* loci in the horse. Animal Blood Groups and Biochemical Genetics 9:169–173.

Schneider, J.E. and Leipold, H.W., 1978. Recessive lethal white in two foals. Journal of Equine Medicine and Surgery 2:479–482.

Scott, A.M., 1970. A single acid gel for the separation of albumins and transferrins in horses. Animal Blood Groups and Biochemical Genetics 1:253–254.

Scott, A.M., 1978. Immunogenetic analysis as a means of identification in horses. *In* Bryans, J.T. and Gerber, H. (Editors), Equine Infectious Diseases. Veterinary Publications, Princeton, pp. 259–268.

Searle, A.G., 1968. Comparative Genetics of Coat Colour in Mammals. Logos Press, London.

Shamis, L.D. and Auer, J., 1985. Complete ulnas and fibulas in a pony foal. Journal of the American Veterinary Medical Association 186:802–804.

Sharp, A.J., Wachtel, S.S. and Benirschke, K., 1980. H-Y antigen in a fertile XY female horse. Journal of Reproduction and Fertility 58:157–160.

Silvers, W.K., 1979. The Coat Colors of Mice. Springer-Verlag, New York.

Smith, A.T., 1977. Lethal white foals in mating of overo spotted horses. Theriogenology 8:303–312.

Speed, J.G., 1958. A cause of malformation of the limbs of Shetland Ponies with a note on its phylogenic significance. British Veterinary Journal 114:18–22.

Spier, S.J., Carlson, G.P., Harrold, D., Bowling, A., Byrns, G. and Bernoco, D., 1993. Genetic study of hyperkalemic periodic paralysis in horses. Journal of the Veterinary Medical Association 202:933–937.

Sponenberg, D.P., 1982. The inheritance of leopard spotting in the Noriker horse. Journal of Heredity 73:357–359.

Sponenberg, D.P., 1990. Dominant curly coat in horses. Génétique, Sélection, Evolution 22:257–260.

Sponenberg, D.P. and Beaver, B.V., 1983. Horse Color. Texas A&M University Press, College Station.

Sponenberg, D.P., Carr, G., Simak, E. and Schwink, K., 1990. The inheritance of the Leopard complex of spotting patterns in horses. Journal of Heredity 81:323–331.

Sponenberg, D.P., Harper, H.T. and Harper, A.L., 1984. Direct evidence for linkage of roan and extension loci in Belgian horses. Journal of Heredity 75:413–414.

Steiss, J.E. and Naylor, J.M., 1986. Episodic muscle tremors in a Quarter Horse: resemblance to Hyperkalemic Periodic Paralysis. Canadian Veterinary Journal 27:332–335.

Stormont, C., 1975. Neonatal isoerythrolysis in domestic animals: a comparative review. Advances in Veterinary Science and Comparative Medicine 19:23–45.

Stormont, C. and Suzuki, Y., 1963. Genetic control of albumin phenotypes in horses. Proceedings of the Society for Exprimental Biology and Medicine 114:673–675.

Stormont, C. and Suzuki, Y., 1964. Genetic systems of blood groups in horses. Genetics 50:915–929.

Sundberg, J.P., Burnstein, T., Page, E.H., Kirkham, W.W. and Robinson, F.R., 1977. Neoplasms of the Equidae. Journal of the American Veterinary Medical Association 170:150–152.

Tassabehji, M., Newton, V.E. and Read, A.P., 1994. Waardenburg syndrome type 2 caused by mutations in the human microphthalmia (MITF) gene. Nature Genetics 8:251–255.

Tolley, E.A., Notter, D.R. and Marlowe, T.J., 1985. A review of the inheritance of racing performance in horses. Animal Breeding Abstracts 53:163–185.

Tozaki, T., Sakagami, M., Mashima, S., Hirota, K. and Mukoyama, H., 1995. ECA-3: equine (CA) repeat polymorphism at chromosome 2p1.3-4. Animal Genetics 26:283.

Traub-Dargatz, J.L., McClure, J.J., Koch, C. and Schlipf, J.W., Jr., 1995. Neonatal isoerythrolysis in mule foals. Journal of the Veterinary Medical Association 206:67–70.

Trommershausen-Smith, A., 1978. Linkage of tobiano coat spotting and albumin markers in a pony family. Journal of Heredity 69:214–216.

Trommershausen-Smith, A., Suzuki, Y. and Stormont, C., 1976a. Use of blood typing to confirm principles of coat-color genetics in horses. Journal of Heredity 67:6–10.

Trommershausen-Smith, A., Suzuki, Y. and Stormont, C., 1976b. Alloantibodies: their role in equine neonatal isoerythrolysis. *In* Proceedings, First International Symposium on Equine Hematology. Michigan State University, May 1975, pp. 349–355.

Troyer, D., Howard, D., Leipold, H.W. and Smith, J.E., 1989. A human minisatellite sequence reveals DNA polymorphism in the equine species. Journal of Veterinary Medicine, A 36:81–83.

Van Haeringen, H., Bowling, A.T., Lenstra, J.A., Zwaagstra, K.A. and Stott, M.L., 1994. A highly polymorphic horse microsatellite locus VHL20. Animal Genetics 25:207.

Van Vleck, L.D., 1990a. Relationships and inbreeding. *In* Evans, J.W., Borton, A., Hintz, H.F. and Van Vleck, L.D. The Horse. W.H. Freeman, New York, pp. 537–554.

Van Vleck, L.D., 1990b. Principles of selection for quantitative traits. *In* Evans, J.W., Borton, A., Hintz, H.F. and Van Vleck, L.D., The Horse. W.H. Freeman, New York, pp. 555–590.

Van Vleck, L.D. and Davitt, M., 1977. Confirmation of a gene for dominant dilution of horse colors. Journal of Heredity 68:280–282.

von Butz, H. and Meyer, H., 1957. Epitheliogenesis imperfecta neonatorum equi. Deutsches Tierarztliche Wochenschrifts 64:555–559.

Vonderfecht, S.L., Bowling, A.T. and Cohen, M., 1983. Congenital intestinal aganglionosis in white foals. Veterinary Pathology 20:65–70.

Watson, A.G. and Mayhew, I.G., 1986. Familial congenital occipitoatlantoaxial malformation (OAAM) in the Arabian horse. Spine 11:334–339.

Weitkamp, L.R., Bailey, E., MacCluer, J.W. and Guttormsen, S.A., 1989. Polymorphism for equine F13A: linkage of F13A with ELA and A. Animal Genetics 20, suppl. 1:10–11.

Weitkamp, L.R., Costello-Leary, P. and Guttormsen, S.A., 1983. Equine marker genes: polymorphism for plasminogen. Animal Blood Groups and Biochemical Genetics 14:219–223.

Weitkamp, L.R., Costello-Leary, P. and Guttormsen, S.A., 1985. Equine marker genes: polymorphism for haptoglobin and assignment of the locus for haptoglobin to equine linkage group II. Animal Blood Groups and Biochemical Genetics 16, Suppl. 1:78–79.

Weitkamp, L.R., Guttormsen, S.A. and Costello-Leary, P., 1982. Equine gene mapping: close linkage between the loci for soluble malic enzyme and Xk (Pa). Animal Blood Groups and Biochemical Genetics 13:279–284.

Whitehouse, D.B., Evans, E.P., Putt, W. and George, A.M., 1984. Karyotypes of the East African common zebra, *Equus burchelli*: centric fission in a pedigree. Cytogenetics and Cell Genetics 38:171–175.

Williams, H., Richards, C.M., Konfortov, B.A., Miller, J.R. and Tucker, E.M., 1993. Synteny mapping the horse using horse–mouse heterohybridomas. Animal Genetics 24:257–260.

Witham, C.L., Carlson, G.P. and Bowling, A.T., 1984. Neonatal isoerythrolysis in foals: management and prevention. California Veterinarian 11:21–34.

Witzel, D.A., Joyce, J.R. and Riis, R.C., 1977. Electroretinography studies of night blindness in the Appaloosa. *In* Proceedings, American College of Veterinary Ophthalmology, pp. 93–100.

Wong, P.L., Nickel, L.S., Bowling, A.T. and Steffey, E.P., 1986. Clinical survey of anti-red blood cell antibodies in horses after homologous blood transfusion. American Journal of Veterinary Research 47:2566–2571.

Woolf, C.M., 1989. Multifactorial inheritance of white facial markings in the Arabian horse. Journal of Heredity 80:173–178.

Woolf, C.M., 1990. Multifactorial inheritance of common white markings in the Arabian horse. Journal of Heredity 81:250–256.

Woolf, C.M., 1991. Common white facial markings in bay and chestnut Arabian horses and their hybrids. Journal of Heredity 82:167–169.

Woolf, C.M. and Swafford, J.R., 1988. Evidence for eumelanin and pheomelanin producing genotypes in the Arabian horse. Journal of Heredity 79:100–106.

Wriedt, C., 1924. Vererbungsfaktoren bei weissen Pferden im Gestüt Frederiksborg. Zeitschrift fürTierzüchtung und Züchtungsbiologie 1:231–242.

Xu, X. and Årnason, Ú, 1994. The complete mitochondrial DNA sequence of the horse, *Equus caballus:* extensive heteroplasmy of the control region. Gene 148:357–362.

Yut, J. and Weitkamp, L.R., 1979. Equine peptidases: correspondence with human peptidases and polymorphism for erythrocyte peptidaseA. Biochemical Genetics 17:987–994.

Zhou, J., Spier, S.J., Beech, J. and Hoffman, E.P., 1994. Pathophysiology of sodium channelopathies: correlation of normal/mutant mRNA ratios with clinical phenotype in dominantly inherited periodic paralysis. Human Molecular Genetics 3:1599–1603.

INDEX

LIBRARY
WARWICKSHIRE
COLLEGE